Rotary Drum

Rotary Drum: Fluid Dynamics, Dimensioning Criteria, and Industrial Applications provides in-depth analysis of fluid dynamics in rotary drums. In addition, it provides analysis on the different configurations, including nonconventional ones, diverse industrial applications, and comparison with competing dryer types, as well as the modeling of these devices.

Covering important aspects of fluid dynamics in rotary drums, which directly influence the drying performance, the book also considers the significant cost of conventional rotary dryers. It takes into account the scale-up of rotary dryers and the control of product quality during processing, which can leave the final product overdried and overheated, wasting thermal energy.

The book serves as a useful reference for researchers, graduate students, and engineers in the field of drying technology.

W0234705

Advances in Drying Science and Technology

Series Editor: Arun S. Mujumdar
McGill University, Quebec, Canada

Handbook of Industrial Drying, Fourth Edition
Arun S. Mujumdar

Advances in Heat Pump-Assisted Drying Technology
Vasile Minea

Computational Fluid Dynamics Simulation of Spray Dryers: An Engineer's Guide
Meng Wai Woo

Handbook of Drying of Vegetables and Vegetable Products
Min Zhang, Bhesh Bhandari, and Zhongxiang Fang

Intermittent and Nonstationary Drying Technologies: Principles and Applications
Azharul Karim and Chung-Lim Law

Thermal and Nonthermal Encapsulation Methods
Magdalini Krokida

Industrial Heat Pump-Assisted Wood Drying
Vasile Minea

Intelligent Control in Drying
Alex Martynenko and Andreas Bück

Drying of Biomass, Biosolids, and Coal: For Efficient Energy Supply and Environmental Benefits
Shusheng Pang, Sankar Bhattacharya, Junjie Yan

Drying and Roasting of Cocoa and Coffee
Ching Lik Hii and Flavio Meira Borem

Heat and Mass Transfer in Drying of Porous Media
Peng Xu, Agus P. Sasmito, and Arun S. Mujumdar

Freeze Drying of Pharmaceutical Products
Davide Fissore, Roberto Pisano, and Antonello Barresi

Frontiers in Spray Drying
Nan Fu, Jie Xiao, Meng Wai Woo, Xiao Dong Chen

Drying in the Dairy Industry
Cécile Le Floch-Fouere, Pierre Schuck, Gaëlle Tanguy, Luca Lanotte, and Romain Jeantet

Spray Drying Encapsulation of Bioactive Materials
Seid Mahdi Jafari and Ali Rashidinejad

Flame Spray Drying: Equipment, Mechanism, and Perspectives
Mariia Sobulska and Ireneusz Zbicinski

For more information about this series, please visit: www.routledge.com/Advances-in-Drying-Science-and-Technology/book-series/CRCADVSCITEC

Rotary Drum
Fluid Dynamics, Dimensioning Criteria, and Industrial Applications

Marcos Antonio de Souza Barrozo
Dyrney Araújo Dos Santos
Cláudio Roberto Duarte
Suellen Mendonça Nascimento

CRC Press
Taylor & Francis Group
Boca Raton London New York

CRC Press is an imprint of the
Taylor & Francis Group, an **informa** business

Designed cover image: Marcos A.S. Barrozo, Dyrney Araújo Dos Santos, Cláudio Roberto Duarte, and Suellen Mendonça.

First edition published 2024
by CRC Press
6000 Broken Sound Parkway NW, Suite 300, Boca Raton, FL 33487-2742

and by CRC Press
4 Park Square, Milton Park, Abingdon, Oxon, OX14 4RN

CRC Press is an imprint of Taylor & Francis Group, LLC

© 2024 Taylor & Francis Group, LLC

Library of Congress Cataloging-in-Publication Data
Names: Barrozo, Marcos A., author. | Santos, Dyrney Araújo dos, author. | Duarte, Cláudio Roberto, author. | Mendonça, Suellen, author.
Title: Rotary drum : fluid dynamics, dimensioning criteria, and industrial applications / Marcos A. Barrozo, Dyrney Araújo Dos Santos, Cláudio Roberto Duarte, Suellen S Mendonça.
Description: First edition. | Boca Raton : CRC Press, 2023. | Series: Advances in drying science and technology, 2475-6229 | Includes bibliographical references and index.
Identifiers: LCCN 2022060348 (print) | LCCN 2022060349 (ebook) |
ISBN 9781032191898 (hardback) | ISBN 9781032192000 (paperback) | ISBN 9781003258087 (ebook)
Subjects: LCSH: Rotary dryers. | Computational fluid dynamics. | Discrete element method.
Classification: LCC TP363 .B385 2023 (print) | LCC TP363 (ebook) |
DDC 620.1/064–dc23/eng/20230406
LC record available at https://lccn.loc.gov/2022060348
LC ebook record available at https://lccn.loc.gov/2022060349

ISBN: 978-1-032-19189-8 (hbk)
ISBN: 978-1-032-19200-0 (pbk)
ISBN: 978-1-003-25808-7 (ebk)

DOI: 10.1201/9781003258087

Typeset in Times
by codeMantra

Contents

Preface

Rotary drums have been used in the industry for many years in several applications, including drying. The large use of these devices is due to their flexibility to handle a wider range of materials than other types of dryers and their high processing capacity. However, despite its simplicity and flexibility, conventional rotary dryer units represent a significant cost to industry. The design and scale-up of rotary dryers remain complex and poorly described. Thus, because of the complexity in estimating some parameters, most rotary dryers are overdesigned and consequently, the final product can be overdried and overheated, wasting thermal energy.

It is important that engineers be able to perform successful designs of these dryers, and it is equally important that they know or be able to understand how the main variables influence their design and performance. The effects of drum filling degree, the physical properties of the material, the drum inclination, the rotational speed, the length and diameter of the cylinder, the number and type of flights, and the residence time on the performance of these devices need to be better understood.

The rotary drum design and prediction of performance are intrinsically related to the ability of estimating all the complex transport phenomena inside this equipment, such as the momentum, including the particle–particle and particle–fluid interactions, and the heat and mass transfer. Besides experimental techniques, numerical simulations can be used to fundamentally investigate the fluid–solid interaction inside rotary drums, in an attempt to connect the material properties and operating conditions with the measured results, consequently removing the empiricism toward predictive design and operation.

Due to some limitations, such as wasting thermal energy and the fact that they not be used to process pasty materials, conventional rotary dryers have undergone some modifications in order to improve their performance and applications.

The objective of the presentation in this book is to impart a knowledge obtained in our research with rotary dryers. In this book, we discuss some important aspects of fluid dynamics in rotary drums, which directly influence the drying performance of these devices; in addition, we also present some important analysis about the different configurations, including nonconventional ones, and comparison with competing dryer types, as well as about the modeling of these devices. Chapter 2 will focus on the description of the flight formats and the geometric studies for the optimal loading estimation and better use of the active drying area of the dryers, while Chapter 3 will address the application of nonconventional dryers, aiming to improve the drying efficiency and preserve the quality and properties of interest of the material to be dried.

The details of computational fluid dynamics (CFD) and discrete element method (DEM) approaches to represent the particle dynamics inside a flighted rotary dryer, as well as the mathematical description of the physical phenomena associated with the drying process, are presented in Chapter 4 of this book.

Uberlândia, November of 2022
Marcos Antonio de Souza Barrozo
Dyrney Araújo Dos Santos
Cláudio Roberto Duarte
Suellen Mendonça Nascimento

Authors

Prof. Marcos Antonio de Souza Barrozo graduated in Chemical Engineering from the Federal University of Rio de Janeiro (UFRJ, 1983), MSc in Chem. Engineering from COPPE/UFRJ (1985), Ph.D. in Chem. Engineering from UFSCar (1995) and Postdoc at the University of British Columbia (UBC) in Canada. He is a Full Professor at Federal University of Uberlândia (UFU). He has experience in chemical engineering, acting on the following topics: fluid dynamics of particulate systems, solid–liquid separation, drying, coating, pelleting, and flotation. He is CNPq Researcher 1A, Member of the Standing Committee of the Brazilian Congress of Particulate Systems. He was coordinator of the Board of Architecture and Engineering (TEC) of FAPEMIG, Member of the Board of Chemical Engineering at CNPq, and coordinator of the Graduate Program in Chemical Engineering of UFU. He has supervised 135 graduate students, 86 MSc students, 49 PhD students, and 14 postdoctoral students. He is the author of 242 papers

Prof. Dyrney Araújo Dos Santos is an Assistant Professor of Chemical Engineering at Federal University of Goiás (UFG), Institute of Chemistry, Brazil, since 2016. He received his bachelor's degree in Chemical Engineering (2009), MSc in Chemical Engineering (2011), and PhD in Chemical Engineering (2015) from the Federal University of Uberlândia (UFU), Brazil. In 2020, he received a postdoctoral fellowship from the Alexander von Humboldt Foundation to conduct research at Friedrich-Alexander University Erlangen-Nuremberg (FAU) in Germany. He is a member (alumni) of the Alexander von Humboldt Foundation and a member of the Brazil Humboldt Club. His research interests include systems involving multiphase fluid-particle flows, e.g., spouted beds, fluidized beds, rotary drums, and fluidized bed opposed gas jet mills, through experiments and numerical simulations using the Euler–Euler and the Euler–Lagrange approaches.

Prof. Cláudio Roberto Duarte is Associate Professor of Chemical Engineering at Federal University of Uberlândia (UFU), Chemical Engineering Faculty, Brazil, since 2006. He received his bachelor's degree in Chemical Engineering (2000), MSc in Chemical Engineering (2002), and PhD in Chemical Engineering (2006) from the Federal University of Uberlândia (UFU), Brazil. His research interests include systems involving multiphase fluid-particle flows, e.g., granulation, mills, spouted beds, rotary drums, and solid and liquid separation, through experiments and numerical simulations using the Euler–Euler and the Euler–Lagrange approaches.

Prof. Suellen Mendonça Nascimento is an Adjunct Professor at Federal University of Lavras (UFLA), Brazil, since 2018. She graduated in Chemical Engineering (2013), MSc in Chemical Engineering (2014), and PhD in Chemical Engineering (2018) from the Federal University of Uberlândia (UFU), Brazil. Her research interests include fluid dynamics of particulate systems, e.g., rotary drums with flights, rotary dryers, and spouted beds, through experiments and numerical simulations using Computational Fluid Dynamics (CFD) and Discrete Element Method (DEM).

1 Introduction

Drying is a liquid and solid separation process by the action of heat that results in evaporation of the liquid. The liquid extracted by evaporation is commonly but not exclusively water. Since the prehistoric times, its application is aimed at food preservation, such as dehydration of fruits, grains, and meats exposed to the sun and wind currents in a process called natural drying. The water content present in solids is called moisture. During the drying process, there is heat and mass transfer, which, in addition to a change in solid moisture, may result in physical and chemical transformations. The heat transfer required to evaporate the surface moisture of solids depends on external conditions of temperature, air humidity, air flow and direction, area of exposure of the solid, etc. Therefore, it is necessary to develop processes that can control such variables.

As production and industrial capacities advance, a variety of techniques and devices are being developed and improved in order to reduce the moisture of materials.

Currently, there is a lot of equipment intended for the drying process involving different technologies and energy sources. Despite the wide application of drying in the processing industry, it is common to associate its success with the practical experience of operators. This is in the opposite direction of the future of Industry 4.0 application, which aims to relate automation and efficiency. The application of information technology, artificial intelligence, and robotics requires the most complete and broad predictive capacity of phase behavior during the drying process.

An excellent option of industrial dryers, rotary dryers stand out due to their simple construction, easy operation, and high processing capacity. However, their evolution based on practical experience has contributed to many inefficient or poorly sized industrial dryers. Although robust and mechanically stable, this type of equipment consumes a high amount of energy. A small increase in the drying efficiency, for instance, could save millions of dollars per year. The main challenge to be overcome in the modeling of mass and heat transfer of a rotary dryer, and consequently its optimized design, refers to the adequate prediction of the granular material dynamics. A satisfactory design requires the knowledge of the dynamics of particles inside the equipment, from the entry to the discharge of the material, as well as the material particularities, the operational condition, and the equipment itself.

Studies on the complexity level, drying, and granular material dynamics are often carried out separately. This allows the designer to make inferences about the actual process, without exactly making a reliable prediction. Therefore, investigations about the operation of rotary dryers are mostly directed to a better understanding of the particle movement along the drum length.

As described by Perry and Green (1997), rotary dryers can be classified as direct, indirect–direct, indirect, and special types. This classification is related to the heat transfer method between the fluid phase and the particulate matter. While in the

DOI: 10.1201/9781003258087-1

direct type, the heat exchange is caused by direct contact between the granular material and the gases, in the indirect type it occurs through indirect contact, i.e., through a wall (usually metallic). This book will deal only with direct rotary dryers, i.e., those in which the contact between the gases and the particulate material occurs directly without the presence of a separating wall.

According to the same authors (Perry and Green, 1997), direct rotary dryers are metal cylinders with or without flights used in operations involving low or medium temperatures, with the temperature being limited by the metal resistance characteristics used in manufacturing. This equipment can be operated in batch or continuous mode and with concurrent or countercurrent flow in relation to the feeding of gases and granular material.

Because of their simplicity, direct rotary dryers require lower cost of construction and operation. They are used when direct contact between solids and gases can be tolerated without causing damage to the desired product. As this type of equipment operates at a velocity greater than 0.5 m/s, very thin particulate matters may result in drag and material loss.

Very common in this type of equipment, the use of flights has the purpose of promoting the drying of free-flowing particulate matter, such as grains, sugar, and ores (Lee, 2008). According to Sheehan et al. (2005), its vast applicability is mainly due to its high processing capacity when compared to other dryers, and flexibility to operate with different types of materials.

Usually, rotary drums consist of a cylindrical drum, with a small horizontal inclination from 0° to 5°, which revolves around its own longitudinal axis, as shown in Figure 1.1.

Coupled internally to the drum, more specifically in its inner circumference, are some flights, which are responsible for promoting the cascade of solids, increasing its contact with the drying air. Their length-to-diameter ratio can vary between 4 and 10, while their diameter can range from 0.2 to 3 m (Perry and Green, 1997). These flights allow particles to be carried and distributed, increasing the contact surface area with the hot gas stream (Geng et al., 2011; Sunkara et al., 2013). The presence of these structures adds greater complexity to the description of the behavior of granular dynamics within the equipment.

The wet particulate material is fed at one end of the drum. The contact of the particles and the dryer wall or the flights promotes the loading and unloading of the

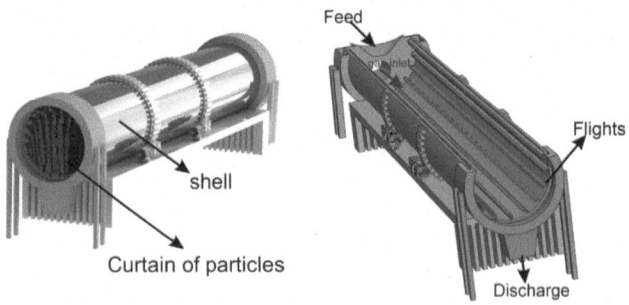

FIGURE 1.1   Direct-heat rotary dryer.

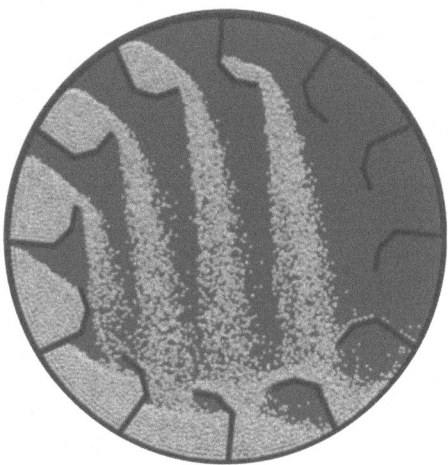

FIGURE 1.2   Curtain of particles formed in the cross-section of the rotary dryer.

latter along a rotation cycle. The flights carry the solids to the top half of the drum, from where they are discharged through a "curtain" of particles formed in the cross-section of the drum, as illustrated in Figure 1.2.

This curtain of particles is crossed by the drying gas, and due to gravity action, it reaches the bottom of the drum where it is again charged by the flights. The fall of cascading solids from the flights and because of the horizontal inclination contributes to the transport of particles in the axial direction of the dryer until the particulate material leaves it at the opposite end to the feed.

Drying air can be fed on the same side as a particulate (concurrent) material (Figure 1.3a) or on the opposite side (countercurrent) (Figure 1.3b). Therefore, the drying gas may favor or delay the transport of solids if fed in the concurrent or countercurrent flow, respectively.

As already mentioned, the flights promote the cascade of the granular material, increasing the rate of heat and mass transfer between the fluid and the solid, which justifies their use and study. Matchett and Baker (1988) defined two phases in the rotary dryer. The first one, called dense phase, is formed when the material is loaded in the flight and at the bottom of the dryer, while the second one, the dilute phase, is formed when the material falls from the flight. According to the authors, the particles spend 90%–95% of their residence time in the dense phase, whereas most part of the drying process occurs when the material is in the dilute phase. Thus, it is necessary a well-distributed cascade along the drum section so as to make the most of the drying area. The dilute region corresponds to the central region of the dryer, where the disused solid material comes into direct contact with the hot gas and the drying occurs in greater magnitude; this region is named active region.

Many authors report that only about 10% of the residence time of particles corresponds to an effective contact time with the drying gas. Therefore, the flights are responsible for the distribution of particulate matter in the cross-chamber section of the bed along a rotation cycle, which is repeated from the feeding of particles to

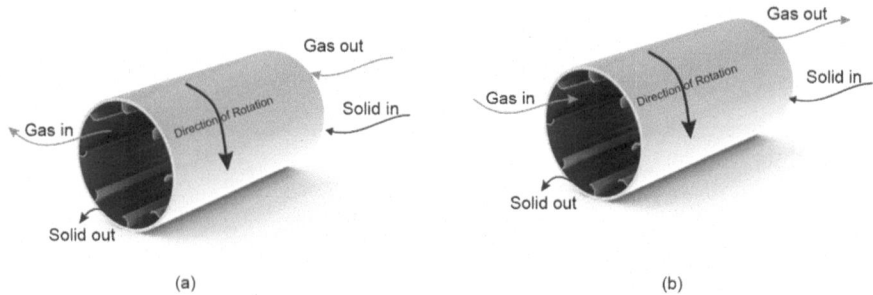

FIGURE 1.3 Flighted rotary dryer for contact between the granular solid and the gas stream in (a) countercurrent and (b) concurrent flows.

their discharge. Different geometries of flights combined with the type of particulate material determine the quality of the granular material curtain. This results in larger or smaller active drying area. Initially, studies focused on the understanding of the particle dynamics in rotary drums were based on geometric models used to predict the loading and unloading profile of a flight.

Some studies in the literature indicate that solids continuously change from one phase to another. In addition, the exchange frequency depends on the properties of solids (i.e., cohesion and dynamic angle of repose) and operational parameters (i.e., air and solid strain rates, drum inclination, rotation frequency, and equipment geometry) (Sheehan et al., 2005; Sherritt et al., 1993).

The description of particle dynamics inside a rotary dryer is of great complexity and requires the understanding and characterization of granular flows as well as the determination of what is a granular material.

Granular material consists of a conglomeration of discrete solids or macroscopic grains. The grain size can vary, depending on the system, and has a characteristic diameter greater than $1 \mu m$. These drying systems are commonly used in the mining, food, storage, pharmaceutical, ceramics, catalyst, and other types of industry, according to the application of interest. These grains or particles are rigid solids with different densities, shapes, and sizes. Although the granular particle individually has a typical behavior of a solid, the set of these particles can behave like a "fluid." It is known that a moving solid has no relative velocity between the particles that compose it. However, the granular material may present distinct velocities, which induces a behavior like that of a fluid. For example, in a sand hourglass (Figure 1.4), the sand is poured from top to bottom, with the particles near the central throat region moving at higher velocity and the particles at the top and near the wall region moving at lower velocities. This is a characteristic of the granular material, which has particular and complex properties for predicting its dynamics.

As seen with the particles at the bottom of the hourglass shown in Figure 1.4, when put on a surface, the granular material reposes following its own dynamics, which depends on moisture and other parameters. In the drying process, the material moisture varies from the input to the output equipment in a continuous operation. A minor change in the material moisture results in a difference in the parameters and

FIGURE 1.4   Sand hourglass.

consequently in the dynamic behavior of the material over its residence time in the dryer. The piling of granular material by deposition on a surface reaches a boundary that is associated with the critical angle of repose. When submitted to a condition that overcomes the critical angle of repose, the material undergoes an avalanche phenomenon (Karali et al., 2015). The avalanche of particles rolling over each other in a pile is referred to as free surface flow, which is of particular interest in the studies of rotary dryers. This avalanche phenomenon occurs throughout the drying process in rotary dryers, being present when the material piles up either on the drum walls or on the flights. During the rotation cycle, the flights change the set of forces that act on the stability of the particulate material set loaded on the flights, inducing a modification in the value of the dynamic angle of repose and promoting an avalanche of the material. When the avalanche occurs, the cascade of the particulate matter during a rotation cycle falls through the cross-section of the bed. Exactly during this free fall from the flights to the base of the bed, the granular material reaches the most effective contact with the drying air.

As already mentioned, flighted rotary dryers are commonly used due to their high processing capacity and application versatility. Their performance depends on the distribution of particles in the cross-section of the drum. The flights are installed along the inner wall of rotary dryers. The rotational movement of the drum makes

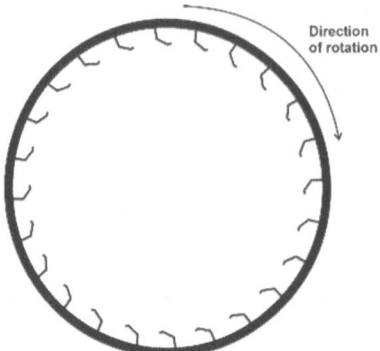 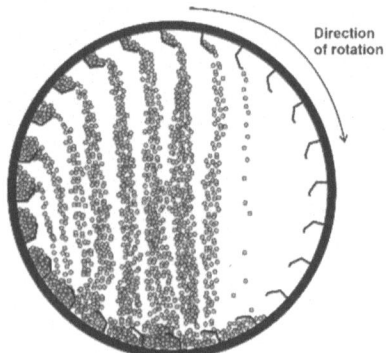

FIGURE 1.5 Typical rotary dryer with flights.

part of the granular material inside the dryer be loaded, forming a static bed on each flight, with a "fluid layer" on the material top. The rotation cycle modifies the inclination of the angle of particles loaded on each flight in relation to the horizontal direction. Thus, when the angle of repose is exceeded, there is the formation of a curtain of particles, enhancing the contact between the particulate matter and the drying air. Ideally, the particle fall is expected to start when the flight is at the 9 o'clock position, as shown in Figure 1.5, which is when it provides the best performance of rotary dryers, offering a better use of the useful drying area. The effective drying area refers to the area occupied by the particulate matter during its fall from the flights to the base of the bed (active region).

The flight shape controls the cascade pattern of particles in the active region where the fluid–particle contact is more effective. Since the design of these flights is a complex task, further studies are necessary to gain an in-depth understanding of the particle dynamics. It is known that the flight characteristics and its positioning on the cylindrical shell affect the behavior of particles in a rotary drum. The main types of flights are shown in Figure 1.6.

Some authors, such as Revol et al. (2001), concluded that the movement of solids in the dryer is influenced by different mechanisms. As the drum rotates, each particle is lifted by the flights and falls from a specific height. In each fall, the particle moves toward the exit due to the dryer inclination. In addition, when the particle reaches the bottom of the drum, it collides with the wall or other particles and mixes with them, causing the particles at the bottom of drum to move forward, rolling over each other. The operation consists of loading and unloading cycles of the flights so that the curtain of particles is continuously renewed. The drying gas favors or hinders the advancement of solids according to the concurrent or countercurrent configuration, respectively.

As reported by Kemp and Oakley (1997), the movement of particles inside a rotary dryer is very complex due to the existence of several forces acting on the particles, causing their movement. The gravitational and drag forces exerted by air can be theoretically predicted, while the rebound and roll of particles can only be experimentally evaluated since they are very dependent on the type of material. The

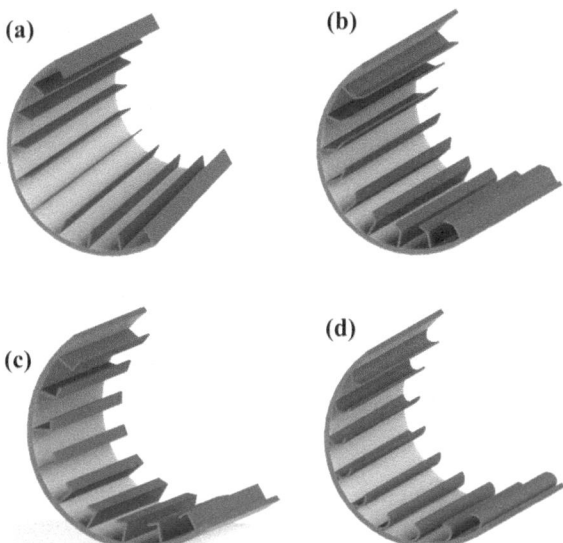

FIGURE 1.6 Main types of flights: (a) straight, (b) angled, (c) right angled, and (d) semicircular.

mathematical modeling of particle dynamics inside drums using the Lagrangian approach is an excellent option for understanding the phenomena involved. However, as a rule, the calibration of parameters strongly depends on the physical properties of the set of particulate matter employed.

Considering the main particle variables associated with the movement and drying rates in a flighted rotary dryer, the length, drum diameter, inclination, rotational speed, particulate matter and gas feed rate, type and quantity of flights, granulometric distribution and other properties of the material stand out. As the performance of these dryers depends mainly on the contact between the solid and the drying gas, fluid dynamics studies are essential to determine the conditions that contribute to increasing the dryer efficiency.

In a rotary dryer, the particles cross regions of variable voidage. In the most dilute region of the flow, that is, the region of greatest voidage, the interaction between the particulate material and the fluid is higher. In regions with less dilute concentrations, there are instantaneous collisions involving several particles, leading to viscous dissipation caused by collisions between particles, the so-called collisional viscous dissipation. On the other hand, in a region of high concentration of solids, i.e., close to the maximum packaging limit (volumetric concentrations of solids greater than 50%), there are no more random oscillations of particles or instantaneous collisions, but intimate and lasting contacts as well as sliding and friction of particles. Besides the previous viscous dissipation, the frictional viscous dissipation that arises from high packing should also be considered. Lun et al. proposed a theory for granular flows as an analogy based on the kinetic theory of gases applied to uniform spherical particles. By computational fluid dynamics (CFD), it is possible to apply this mathematical modeling and obtain the corresponding numerical solution and results of interest for a better understanding of the phenomena involved. The CFD technique is detailed in Chapter 4 and represents an important tool for the study and optimization of this type of equipment.

The determination of the drying kinetics is fundamental for the mathematical modeling, correct design, and operation of convective dryers. Furthermore, it shows the predominant mechanism in the mass transfer from the material to the fluid and the corresponding mathematical equations. However, mathematical models of drying kinetics often based on thin-layer drying are not sufficient for the design and control of industrial processes. Most dryers involve a complex fluid dynamics of the fluid phase and the particulate matter, which requires a multiphase mathematical modeling that is difficult to apply to complex systems, such as moving beds and rotary dryers.

CFD models have greatly contributed to the study on the coupling and mass transfer of fluid dynamics. On the other hand, the discrete element method (DEM) approach has enabled a better understanding of the granular material dynamics, which is essential for the design and development of moving bed systems. Nevertheless, both approaches are limited to the processing and storage capacity of computers. Despite the great advance in the area in recent years, the number of studies focusing on the CFD-DEM approach to predict the dynamics of the phases involved and the simultaneous heat and mass transfer is still incipient. However, it is true that the DEM approach alone has considerably enhanced the understanding of the granular material dynamics, which is fundamental for the design and development of moving bed systems, for instance. This challenge becomes even greater if the scaling up of industrial dryers is considered.

Widely used for particle dynamics studies, the DEM proposed by Cundall and Strack (1979) models the dynamics between particles in a granular system. To describe the rotational and translational movements of particles, the DEM employs Newton's second law in its differential form. In this method, the trajectory of each particle is monitored considering its interactions with neighboring particles and the equipment walls. The details of the DEM approach are presented in this book to represent the particle dynamics inside a flighted rotary dryer.

The mathematical description of the physical phenomena associated with both the drying process and the development of models capable of predicting the drying kinetics of some material has contributed enormously to the understanding of the phenomena involved as well as to the improvement of processes.

REFERENCES

Cundall, P.A.; Strack, O.D.L. "A discrete numerical model for granular assemblies" *Geotechnique* 29 (1979): 47–65. https://doi.org/10.1680/geot.1979.29.1.47.

Geng, F.; Li, Y.; Wang, X.; Yuan, Z.; Yan, Y.; Luo, D. "Simulation of dynamic processes on flexible filamentous particles in the transverse section of a rotary dryer and its comparison with ideo-imaging experiments" *Powder Technology* 207 (2011): 175–182. https://doi.org/10.1016/j.powtec.2010.10.027.

Karali, M.A. Analysis study of the axial transport and heat transfer of a flighted rotary drum operated at optimum loading. Ph.D. thesis, Otto Von Guericke University (2015), Germany.

Kemp, I.C.; Oakley, D.E. "Simulation and scale-up of pneumatic conveying and cascading rotary dryers" *Drying Technology* 15 (1997): 1699–1710. https://doi.org/10.1080/07373939708917319.

Lee, A. Modelling the solids transport phenomena within flighted rotary dryers. Ph.D. thesis, James Cook University (2008), Australia.

Matchett, A.J.; Baker, C.G.J. "Particle residence times in cascading rotary dryers. Part 2: Application of the two-stream model to experimental and industrial data" *Journal of Separation Process Technology* 9 (1988): 5–13.

Perry, R.H.; Green, D.W. *Perry's Chemical Engineers' Handbook*, 7th ed., McGraw-Hill, New York, 1997.

Revol, D.; Briens, C.L.; Chabagno, J.M. "The design of flights in rotary dryers" *Powder Technology* 121(2001): 230–238. https://doi.org/10.1016/S0032-5910(01)00362-X.

Sheehan, M.E.; Britton, P.F.; Schneider, P.A. "A model for solids transport in flighted rotary dryers based on physical considerations" *Chemical Engineering Science* 60 (2005): 4171–4182. https://doi.org/10.1016/j.ces.2005.02.055.

Sherritt, R.G.; Caple, R.; Behie, L.A.; Mehrotra, A.K. "The movement of solids through flighted rotating drums: Part I model formulation" *The Canadian Journal of Chemical Engineering* 71 (1993): 337–346. https://doi.org/10.1002/cjce.5450710302.

Sunkara, K.R.; Herz, F.; Specht, E.; Mellmann, J. "Influence of flight design on the particle distribution of a flighted rotating drum" *Chemical Engineering Science* 90 (2013): 101–109. https://doi.org/10.1016/j.ces.2012.12.035.

2 Direct-Heat Rotary Dryers

2.1 INTRODUCTION

Among the variables that influence the design and efficiency of rotary dryers are the drum filling degree, the moisture content of solids, the drying air speed, the physical properties of the material, the drum inclination, the rotational speed, the length and diameter of the cylinder, the number and type of flights, and the residence time (Konidis, 1984). Capacity and dimensions are chosen according to the equipment application.

2.2 DIMENSIONS AND DESIGN OF ROTARY DRYERS

The cylinder diameter of rotary dryers can vary from 0.3 to 3 m, which guarantees applicability from bench scale to large industries. The length-to-diameter ratio is between 4 and 10. The inclination, which helps in the transport of the particulate material, ranges from 0 to 0.8 cm/m and can be negative in cases of countercurrent air flow.

Inside the rotating cylinders, there may be flights, as shown in Figure 2.1, with sizes between 1/12 and 1/8 of the drum diameter responsible for carrying the particles and distributing them over the drying gas stream, creating curtains of particles. These curtains can be continuous along the entire drum length, or in order to be more evenly distributed over the gas stream, every 0.6–2 m the flights can be shifted in relation to the previous radial position.

The design of the flights is of utmost importance since one of the factors that influence the drying rate is how the contact between the solid material and the drying gas occurs. Depending on the shape, size, and number of flights in the equipment, different amounts of solids form the falling curtains of particles and make up the active phase of the drum where the drying takes place.

2.2.1 ACTIVE AND PASSIVE PHASES

Matchett and Baker (1988) classified the particles in rotary dryers into two distinct phases: the airborne phase, formed when the curtains of particles fall from the flights, and the dense phase (flight-borne phase), formed when the material is carried by the flights and the particles roll on the bottom of the equipment. This classification helps in the equipment modeling, as the drying occurs practically only when the material is in the airborne phase, giving rise to the term active phase, as opposed to the flight-borne or passive phase. The passive and passive phases are illustrated in Figure 2.2.

DOI: 10.1201/9781003258087-2

FIGURE 2.1   Flights in a rotary dryer (Silveira, 2022).

Although the drying process occurs in the airborne phase, according to Ajayi and Sheehan (2012), the retention time of solids in the active phase is between 10% and 15% of the total retention time and is normally considered invariant in relation to the loading state. Thus, it is necessary that the flight promotes a well-distributed discharge of solids along the drum cross-section in order to make the most of the drying area.

2.2.2 Types of Flights

Given the importance of choosing the flights so as to maximize the retention time of solids in the active region, the main types of flights are shown in Figure 2.3.

The three most used types of flights, illustrated in Figure 2.3, are chosen according to the material used and the position within the equipment.

a. Straight or radial flights:
   They are indicated for very cohesive materials with low flowability, such as wet materials at the beginning of the dryer or cases in which there is a need to increase the residence time in the dryer since these materials do not promote a good distribution of solids over the drying gas.

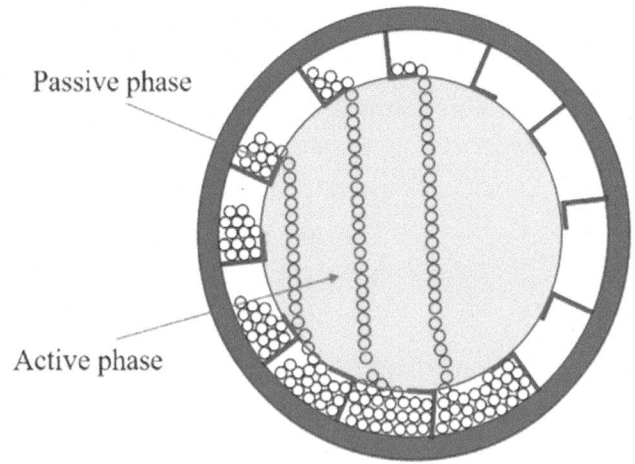

FIGURE 2.2   Active and passive phases in a rotary drum.

b. Three-segmented flights:

This type of flight promotes a good distribution of the material over the active region. Used for free-flowing materials, the three segments provide control over the rate of discharge from the flights, controlling the angle at which the discharge is completed.

c. Two-segmented flights:

For cohesive materials, these flights are more suitable than the three-segmented ones, since the previous type could be a limiting factor for the discharge. When $\alpha_2 = 90°$, this flight is particularly called rectangular flight.

In addition to the number of segments, the angle between one segment and another also influences the discharge of solids, changing the mass of solids in the active region (i.e., the occupation and dispersion of solids in the region) and the angle of last discharge. The addition of the third segment increases the angle of last discharge for free-flowing materials and leads to a better distribution of solids in the active region (Silveira et al., 2020).

Silveira et al. (2020) evaluated the effect of the angles between the segments for flights with two and three segments for free-flowing materials (glass beads). For two-segmented flights, the angle α_2 varied from 80° to 130°. The authors analyzed the mass of solids in the active phase, the area percentage of the active region occupied by the particles and the dispersion heterogeneity of solids in the active region. A better fluid–particle contact was observed for a lower heterogeneity due to the more homogeneous dispersion. A higher α_2 (130°) led to a heterogeneous dispersion, making part of the active region unusable since the flight finished its discharge in smaller angular positions and the fluid passed through empty regions. Smaller α_2 (80° and 90°) led to more homogeneous dispersion, but lower was the mass of solids in the active phase and the area occupied by the particles in the active region.

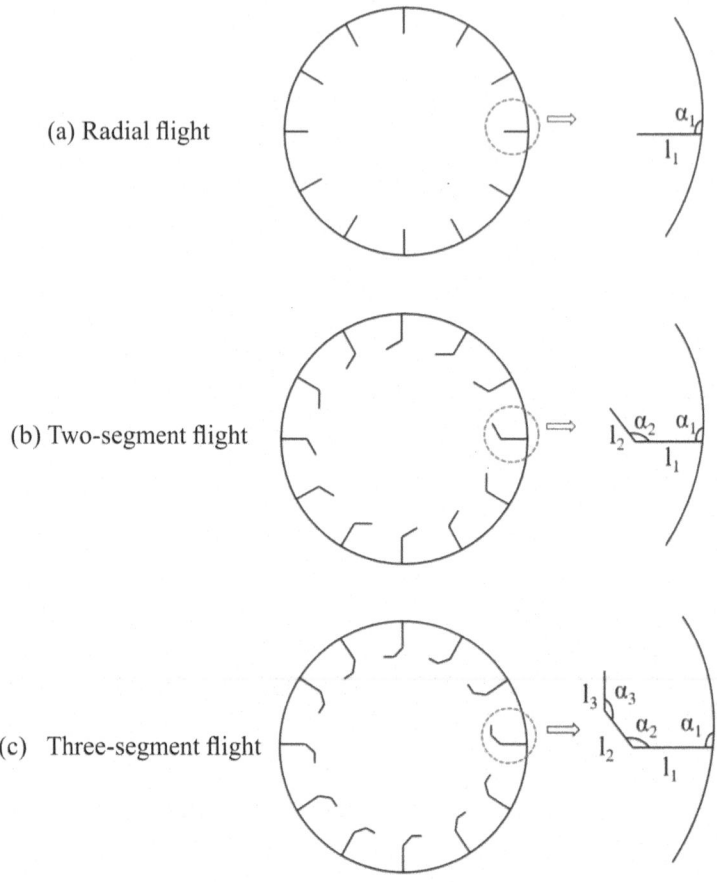

FIGURE 2.3   Main types of flights (Silveira, 2022).

By varying α_3 in three-segmented flights, it could be observed that a higher amount of mass was loaded by the flight. The authors increased α_3 from 115° to 135° and verified that the solids holdup increased to 39.7%. However, a smaller increase of solids holdup occurred for angles greater than 145°. The angular position at which the flight discharge finished was also influenced by α_3. By increasing α_3, the flights finished their discharge in smaller angular positions and part of the active region became empty.

Besides the already mentioned flights, two theoretical models were presented by Kelly (1968), as illustrated in Figure 2.4.

EAD (Equal Angular Distribution) flights are designed for uniform angular distribution. The solids are uniformly discharged over the horizontal plane that passes through the center of the rotary drum, ensuring that the particle distribution is the same between any two equally spaced positions, and consequently guaranteeing a

FIGURE 2.4 Theoretical flights presented by Kelly (1968): (a) EAD – Equal Angular Distance and (b) CBD – Centrally Biased Distribution.

good gas–solid contact in the dryer. In contrast, CBD (Centrally Biased Distribution) flights are intended to ensure that the maximum distribution of particles occurs in the central part of the equipment, which would be the region with the greatest contact with the drying air. The dimensions and angles between the segments depend on the physical characteristics of the particulate material. However, the application of these flights is limited in industries, because for the flow of wet and cohesive materials, the falling curtains of particles are not formed.

According to Konidis (1984), the drying efficiency can be increased in particular cases with the use of different flights in the same dryer. Either straight or two-segmented flights can be used close to the feed when the material is wet and cohesive, while three-segmented flights can be used near the discharge when the material is dry enough to have a free flow.

2.2.3 MOVEMENT OF SOLIDS INSIDE ROTARY DRYERS

The movement of solids inside rotary dryers is influenced by the cylinder rotation, the equipment inclination, and the drag caused by the drying gas flow. This makes the description of the movement of particles complex, as it is subject to the action of centrifugal, gravitational, and frictional forces. The particles go through a set of motions: lifting by the flights and fall through the air stream where they can slide, roll, or bounce at the bottom of the drum, as shown in Figure 2.5.

When the solids enter the equipment, part of them is transported by the flights to the upper half of the cylinder and then fall from these structures, returning to the lower half of the drum. This discharge of solids from the flights forms curtains of particles that fall through the drying gas stream. During the fall, the gas can help or delay the advance of particles, which is always boosted by the equipment inclination.

When the material passes through the flight and is not loaded, it slides on the bottom of the drum by a motion classified as kiln action. Although the material advances in the equipment due to the inclination, it does not pass through the drying

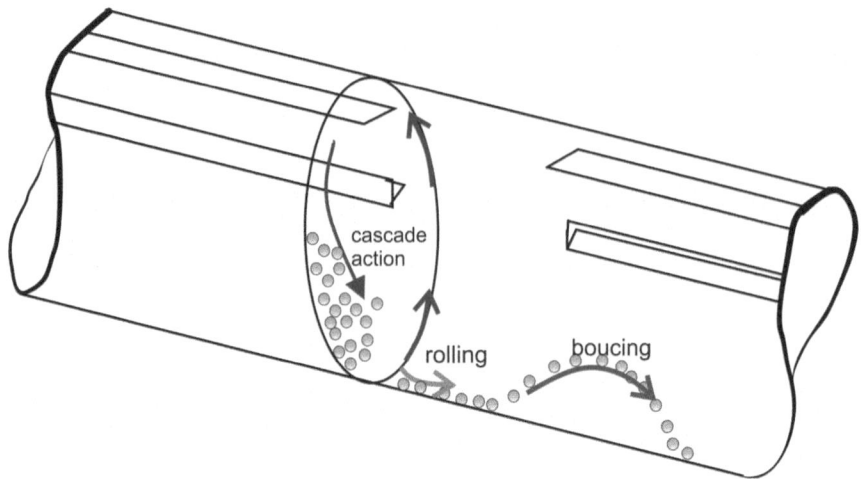

FIGURE 2.5   Movement of particles inside rotary dryers. (Modified from Baker, 1983.)

gas stream, which decreased the product quality. The particles that fall from the flights can collide with the bottom of the bed, another flight or the cylindrical shell, and can roll or bounce.

What determines whether a particle will remain in the flight or be discharged is the dynamic angle of repose (ϕ), which is the angle formed between the surface of the material in the flight and the horizontal plane, as displayed in Figure 2.6.

Schofield and Glikin (1962) apud Mujumdar et al. (2007) presented a model to determine the dynamic angle of repose through a balance among the gravitational, centrifugal, and frictional forces acting on a particle that is about to fall from a suspensor, resulting in Equation (2.1), for v between 0.0025 and 0.04:

$$\tan\phi = \frac{\gamma + v\left(\cos\theta - \gamma\sin\theta\right)}{1 - v\left(\sin\theta + \gamma\cos\theta\right)} \tag{2.1}$$

where θ is the angle between the flight tip and the center of the drum, γ is the coefficient of dynamic friction of the material, and $v = \dfrac{r_H\,\omega^2}{g}$ is the ratio between the centrifugal and gravitational forces acting on the particle, calculated using the radius of the circumference described by the flight tip (r_H).

2.2.4   SOLIDS HOLDUP IN THE FLIGHTS

Concerning the movement of particles inside rotary dryers, it becomes essential to know the mass of solids transported by the flights in order to ensure that the drum is operating in near-optimal conditions. For design-loaded rotary drums, the flights hold the maximum quantity of particles at the 0° angular position. As the drum rotates, the particles are discharged from the flights.

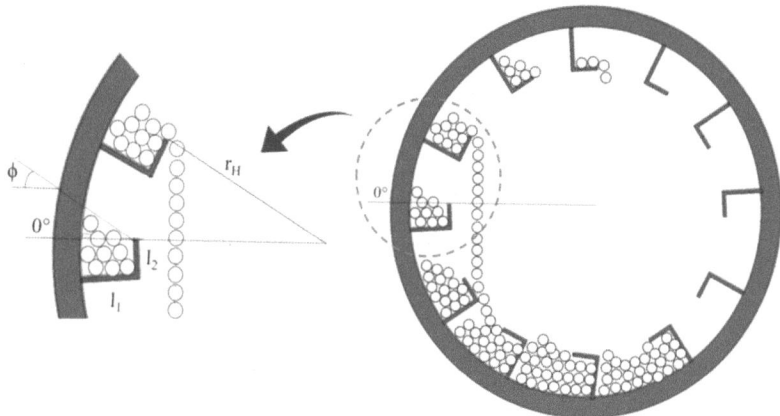

FIGURE 2.6 Flight load in the first unloading quadrant (Silveira, 2020).

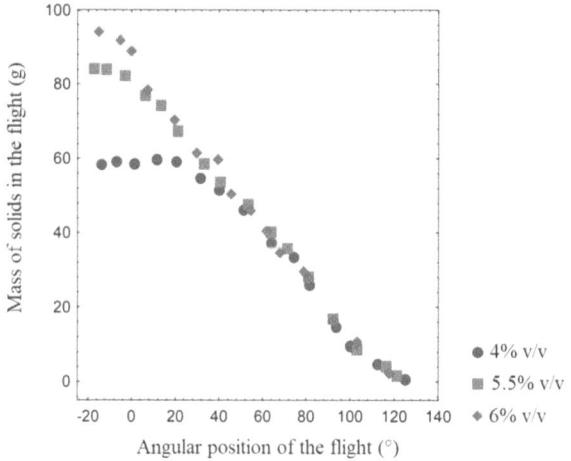

FIGURE 2.7 Mass of solids in the flight as a function of angular position under three different loading conditions of the drum.

Figure 2.7 shows the solids holdup in the flights for the drum operating under three different loading conditions: 4% v/v (underloaded), 5.5% v/v (design-loaded), and 6% v/v (overloaded).

It is observed that the mass of solids in the flight for an underloaded drum remains constant until it saturates and the discharge starts. For a design-loaded rotary drum, the saturation occurs at the 0° angular position, while for an overloaded drum, the discharge of the flights starts before this angular position. The loading conditions will be better explored in Section 2.3.

The mass of solids retained in each flight is dependent on the flight geometry, the physical properties of the particles (i.e., density, dynamic angle of repose, and coefficient of dynamic friction), the rotational speed of the drum and the angular position of the flight.

Figure 2.8 shows the volume occupied by the solids in two flights of different shapes. As it can be seen, three-segmented flights usually have higher internal volume than two-segmented ones. Besides the internal volume of the flight, the mass of solids depends on the dynamic angle of repose (ϕ), that is, the angle formed between the line drawn on the surface of the particles and the horizontal plane. The wider the dynamic angle of repose, the larger the volume of solids in the flight and the greater the mass of solids as a function of the angular position.

The addition of the third segment retards the discharge from the flights, and for free-flowing materials, the distribution over the drying air tends to be better and the discharge tends to finish at higher angular positions.

Besides the type of the flight, the rotational speed of the drum also influences the mass of solids in the flight. Figure 2.9 displays the mass of solids in a two-segmented flight for a design-loaded drum operating with filtered sand at three rotational speeds.

It is observed that the higher the rotational speeds, the greater the solids holdup in the flight as a function of the angular position. In addition, higher rotational speeds lead to greater centrifugal forces acting on the particles, thus retaining particles at higher angular positions in the flights.

The measurement of the mass of solids in the flights must be made at steady state and as a function of the area occupied by the solids, which is the volumetric holdup. If these measurements are made after a sudden stop of the drum, the angle of repose of the particles will change, consequently altering the mass of solids in the flight (Baker, 1988). Thus, the use of photographs to estimate the mass of solids in experimental work is widely used (Revol et al., 2001).

However, since this measurement – especially in industrial scale – is not trivial, some authors have developed geometric models to estimate the flight holdup, which will be explored in the next section.

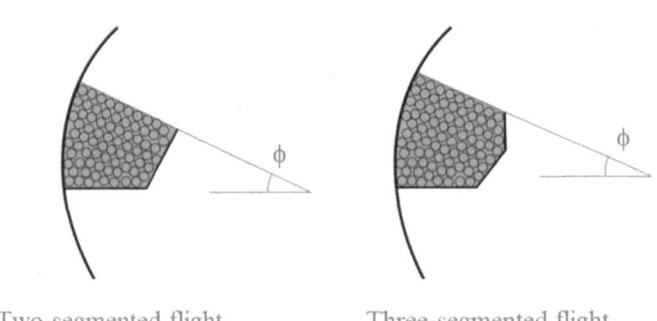

Two-segmented flight   Three-segmented flight

FIGURE 2.8   Volume of solids inside flights of different shapes.

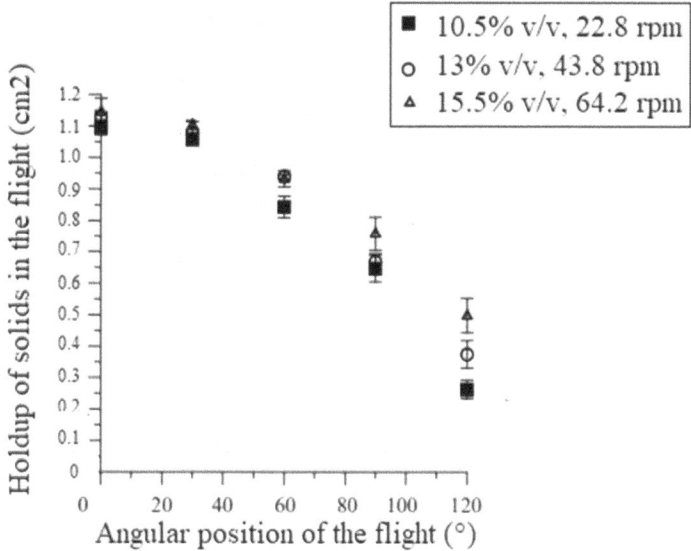

FIGURE 2.9 Discharge of solids from the flights at different rotational speeds.

2.2.5 GEOMETRIC EQUATIONS TO ESTIMATE THE FLIGHT HOLDUP

Revol et al. (2001) proposed an estimate of the coefficient of dynamic friction of a powder to determine the solids holdup in the flights using a theoretical model for three-segmented flights. Lisboa et al. (2007) used the same methodology but for two-segmented flights. Figure 2.10 shows a flight with two segments and an angle of α_A between them.

To calculate the area occupied by the solids in the flights, first, it is necessary to estimate the coordinates of points A and B (Figure 2.10). The angle between the coordinate plane in the center of the drum and the that in the flight tip is called δ.

The equations of the two segments are shown in Equations (2.2) and (2.3):

- Segment 1:

$$y_1 = 0 \tag{2.2}$$

- Segment 2:

$$y_2 = a_2 + b_2 x \tag{2.3}$$

where $a_2 = x_A \tan(\alpha_A)$ and $b_2 = -\tan(\alpha_A)$.

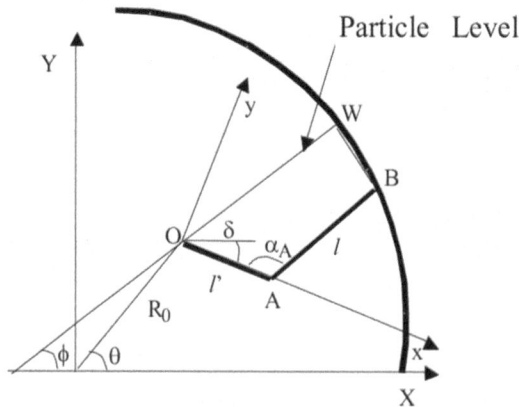

FIGURE 2.10   Scheme of a two-segmented flight (Lisboa et al., 2007).

The coordinates of points A and B are given by:

- point A:

$$x_A = l'' \quad \text{and} \quad y_A = 0 \tag{2.4}$$

- point B:

$$x_B = x_A - l\cos(\alpha_A) \quad \text{and} \quad y_B = l\sin(\alpha_A) \tag{2.5}$$

In the traditional set of coordinates, which is stationary, the position of point B must satisfy Equation (2.6) since it is located on the wall of the drum of radius R:

$$X_B^2 + Y_B^2 = R^2 \tag{2.6}$$

The two coordinate planes are related by Equations (2.7) and (2.8):

$$X_B = X_O + x_B \cos(\delta) - y_B \sin(\delta) = R_O \cos(\theta) + x_B \cos(\delta) - y_B \sin(\delta) \tag{2.7}$$

$$Y_B = Y_O + y_B \cos(\delta) + x_B \sin(\delta) = R_O \sin(\theta) + y_B \cos(\delta) + x_B \sin(\delta) \tag{2.8}$$

The substitution of Equations (2.7) and (2.8) by Equation (2.6) yields Equation (2.9), which estimates the powder level line that can be solved for δ at any angular position:

$$y = x \tan(\gamma) = x \tan(\phi - \delta) \tag{2.9}$$

Its intersection with the line tracing the second segment has the abscissa calculated by Equation (2.10) and the ordinate calculated by Equation (2.11):

$$x_2 = \frac{a_2}{\tan(\gamma) - b_2} . \tag{2.10}$$

$$y_2 = a_2 + b_2 x_2 \tag{2.11}$$

The intersection of the solid level line with the drum wall has the abscissa estimated by Equation (2.12):

$$x_w = -\frac{B_w \pm \sqrt{B_w^2 - 4 A_w C_w}}{2 A_w} \tag{2.12}$$

where $A_w = 1 + \left[\tan(\delta) \right]^2$; $B_w = 2 X_O \left[\cos(\delta) - \tan(\gamma) \sin(\delta) \right] + 2 Y_O \left[\tan(\gamma) \cos(\delta) + \sin(\delta) \right]$ and $C_w = R_O^2 - R^2$. The ordinate is calculated by Equation (2.13):

$$y_w = x_w \tan(\gamma) \tag{2.13}$$

Finally, the area S occupied by the solids in the flight can be calculated in three different situations:

a. The powder reaches the wall. In this situation $\gamma > \arctan\left(\dfrac{y_B}{x_B}\right)$ and the area is calculated by Equation (2.14):

$$S = \frac{R^2}{2} \left[\beta - \sin(\beta) \right] + \frac{1}{2} \left| x_A y_B - x_B y_A + x_B y_w - x_w y_B \right| \tag{2.14}$$

$$\text{where } \beta = 2 \arcsin \left[\frac{\sqrt{(x_B - x_w)^2 + (y_B - y_w)^2}}{2R} \right].$$

b. The powder does not reach the wall, but the second segment. This will occur if $\gamma > \arctan\left(\dfrac{y_B}{x_B}\right)$ and $\sqrt{(x_2 - x_A)^2 + (y_2 - y_A)^2} < l''$. The area is estimated by Equation (2.15):

$$S = \frac{1}{2}|x_A y_2| \qquad\qquad (2.15)$$

c. The flight is empty and $S = 0$. This situation occurs if $y_2 < 0$.

Lisboa et al. (2007) used Equations (2.2)–(2.15) to predict the solids holdup in the flights of a rotary dryer applied to fertilizer drying. Figure 2.11 shows a comparison between the holdup measured experimentally for each angular position of the flight and the results obtained with the proposed equations for the prediction of solids holdup in two-segmented flights.

2.2.6 GEOMETRIC EQUATIONS TO ESTIMATE THE HEIGHT OF FALLING CURTAINS

Since the material to be dried leaves the flights at different angular positions, a range of heights of falling curtains can be observed. If the flights transport the particulate matter to maximize the mean height of the falling curtains, the contact time between them and the drying gas tends to be longer, thus enhancing the heat transfer coefficient and providing a higher drying rate (Silvério et al., 2015).

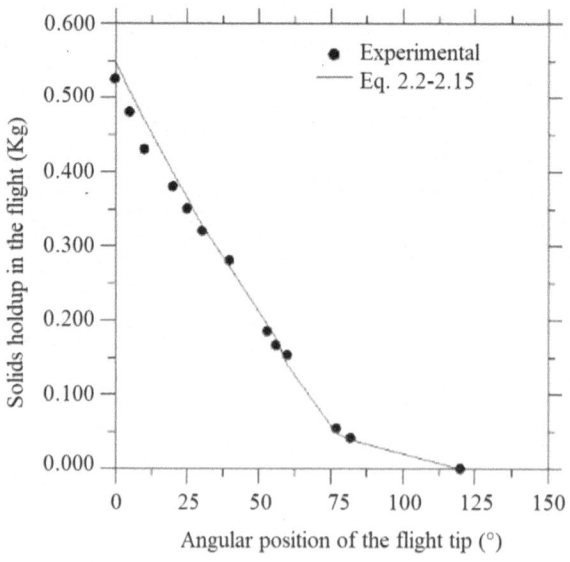

FIGURE 2.11 Geometric determination of the mass of solids in the flight (Lisboa et al., 2007).

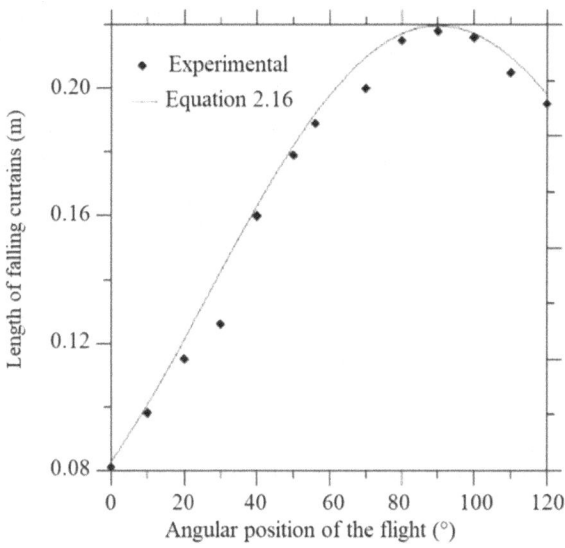

FIGURE 2.12 Geometric determination of the length of falling curtains as a function of the angular position of the flight (Lisboa et al., 2007).

Glikin (1978) proposed Equation (2.16) for calculating the height of falling curtains, with the height being the distance between the flight tip and the particle bed at the bottom of the drum.

$$Y_d = \frac{Y_O + \sqrt{R^2 - X_O^2}}{\cos \alpha} \tag{2.16}$$

where $Y_O = R_0 \cos \theta$ and $X_O = R_0 \sin \theta$.

Lisboa et al. (2007) compared the height of falling curtains obtained by Equation (2.16) and experimental data. The authors observed that such equation provided accurate estimates of these parameters for two-segmented flights, as seen in Figure 2.12.

2.3 DESIGN-LOADED ROTARY DRYERS

2.3.1 DEFINITION

The thermal efficiency of rotary dryers varies between 30% and 60% (Mujumdar, 2014). In order to increase it, there are two important key properties: the residence time, which is the time when the material to be processed is inside the equipment; and the volumetric holdup, characterized by the volume of material in the airborne and dense phases per unit of equipment length, causing the holdup to have an area dimension.

By using an optimal volumetric holdup, the efficiency of rotary dryers can be increased. Typically, this ideal volumetric holdup ranges from 10% to 15% of the

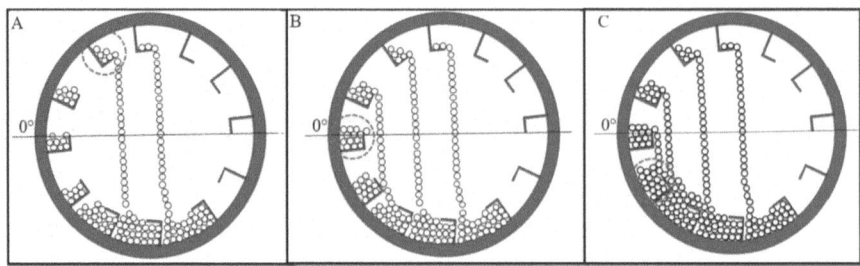

FIGURE 2.13 Rotary drum under three different loading conditions: (a) underloaded, (b) design-loaded, and (c) overloaded (Silveira, 2022).

drum volume filled with solid material to be dried, guaranteeing both equipment efficiency and quality of the dry product (Perry and Green, 1997). To this end, the solids discharge from the flights must start exactly at 0° (Figure 2.13).

Therefore, when the flight passes through 0° without starting to unload, it is operating below its full capacity. The drum is then classified as underloaded. As it can be seen in Figure 2.13, in this loading condition no curtain of solids is formed in a part of the upper half of the drum. Since drying effectively occurs only in the curtains, in this condition the particles have minimal contact with the air, resulting in an underutilization of the equipment useful area. In addition, a short time in the airborne phase can lead to shorter residence times than necessary to reach the desired humidity, decreasing the dryer efficiency (Karali et al., 2015; Ajayi and Sheehan, 2012).

The design load, which would be the ideal load, is the condition in which the flights are operating at their maximum capacity and the unloading starts precisely at the 0° angular position. In this condition, the interaction between the drying gas and the airborne phase is maximum. There are no regions of the upper half of the drum without the formation of curtains of solids and no excess material rolling on the bottom of the equipment, as occurs when the drum is overloaded.

When the quantity of solids is greater than the capacity of the flights, the discharge starts before the solids enter the upper half of the drum and the equipment is overloaded. It is important to note that under these conditions the amount of solids in the flights and curtains is the same as the ideal load. However, the excess material that is not carried by the flights rolls on the base of the drum without effective contact with the drying air. As a consequence, the equipment does not effectively remove moisture from the material as expected, leading to a nonuniformity of the product.

2.3.2 Estimation of the Design Load

Preliminary studies to determine the design load in rotary dryers were developed by Matchett and Baker in 1988. The authors proposed an experimental technique to determine the transition from an underloaded to ideally loaded drum operation. They plotted the dryer holdup as a function of the feed rate and observed a linear behavior of the formed curve. Nonetheless, at a given moment there was a change in the slope of this curve, which evidenced the transition from the underloaded to overloaded condition. According to the authors, the point at which this change occurred in the slope was the design load, suggesting the saturation of the mass of solids in the airborne phase.

Based on the concept of saturation developed by Matchett and Baker (1988), Ajayi (2011) presented four different ways to determine the ideal load:

a. Visual analysis method:

In this method, photographs of the flights are analyzed as they pass through the 0° position – also called the 9 o'clock position. Nascimento et al. (2018) used this technique with a high-speed video camera that allowed to capture the slow-motion movement of the solids in the flights and precisely determine the condition in which the solids began to unload exactly at the 0° position, as shown in Figure 2.14.

According to Ajayi (2011), the main disadvantage of this method is that for materials with high humidity, the discharge may not be constant, and some avalanches may be observed in the falling curtains, which can cause measurement inaccuracy.

b. Change of gradient of total flight-borne solids:

The total flight-borne solids are measured through their holdup in the dense phase, both in the upper and the lower halves of the drum. When the load in the equipment is increased, there is a proportional increase in the solids holdup in the dense phase. However, such proportionality changes from the design loading point, causing a change in the slope of the curve.

The main disadvantage of this method is that this change is not very pronounced, making it difficult its visualization through graphs.

c. Saturation of the airborne phase and solids holdup in the upper half of the drum:

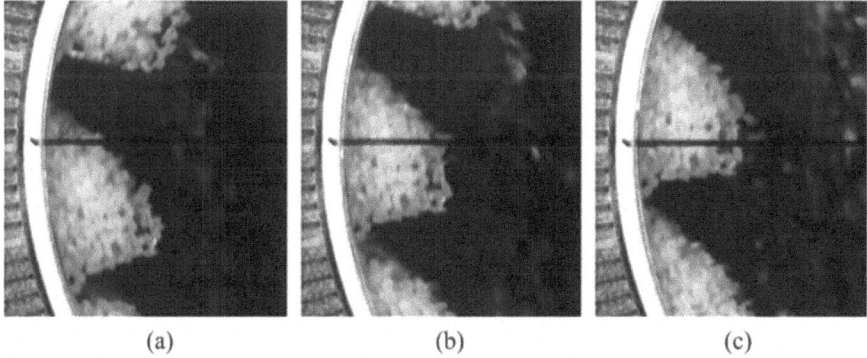

(a)   (b)   (c)

FIGURE 2.14 Visual determination of the design load with the aid of a high-speed camera: (a) before the 0° position, there is no discharge of solids; (b) at the 0° position, the first particles begin to be discharged; and (c) immediately after the position 0°, the curtain of particles starts to form.

In design-loaded drums, the flights carry as many solids as possible, and the airborne phase is occupied by the largest number of particles for this configuration. As the number of particles increases and the drum becomes overloaded, it is not possible to increase the mass of solids either in the flights or in the curtains. Therefore, it can be inferred that in the design load condition, there is saturation of the airborne phase and the flights in the upper half of the drum.

Based on this concept and according to Ajayi (2011), when plotting the graph of the area occupied by the solids in the flights in the upper half of the drum, it is possible to observe an increase in the area with increasing the load until the design load point; after this point, the area becomes constant. A similar behavior is observed for the area occupied by the solids in the airborne phase.

The main disadvantage of this methodology is the fact that there are high standard deviations when measuring the area occupied by both the solids in the upper half of the drum and the curtains of particles due to the difficulty in determining the porosity of the airborne phase, which can propagate errors.

d. Saturation of the flight at the 0° angular position – First Unloading Flight (FUF):

The mass of solids in the dense phase is constant when the drum is overloaded. Therefore, when measuring the area occupied by solids in the flights at 0°, it tends to remain constant after the design load point even with an increase in the drum holdup.

Ajayi (2011) states that before the design load point, the variation in the area occupied by solids in the flights at 0° with increasing the drum holdup is linear and can be adjusted by a first-degree equation that does not necessarily pass through the origin. After this condition, even if the drum holdup increases the load remains constant, allowing these points to be adjusted by a straight line with zero slope. The point of intersection between these two curves is the ideal load point and should be the design load of the equipment.

Despite the different ways used to determine the design load, it is important to develop models that can predict it. In the literature, several authors have used the geometry of the cross-section of flights to propose models that predict the ideal load of the drum, the so-called geometric models.

2.3.3 Geometric Models to Predict the Design Load of Rotary Drums

The first geometric model was developed by Porter (1963), who based the estimation of the dryer design holdup (H_{TOT}) on the holdup of the material in the flight when at the 0° position (h_{FUF}), the so-called First Unloading Flight (FUF). According to the author, for flights in the upper half of the drum, one flight is complementary to another in the opposite position in relation to the vertical line that divides the drum in half for its complete filling. Thus, in the design load condition, the solids are sufficient to fill half of the flights, as shown in Equation (2.17):

$$H_{\text{TOT}} = h_{\text{FUF}} \frac{n_F}{2} \qquad (2.17)$$

where H_{TOT} is the design load holdup, h_{FUF} is the solids holdup in the flight at the 0° position, and n_F is the number of flights in the equipment.

This assumption, however, was not validated in the photographic tests carried out by Matchett and Sheikh (1990), who showed that the mass of solids as a function of the angular position depends on the flight shape, evidencing a lack of universality of the equation.

For example, for flights of EAD, Kelly and O'Donnell (1977) presented the model of Equation (2.18):

$$H_{\text{TOT}} = h_{\text{FUF}} \frac{n_F + 1}{2} \qquad (2.18)$$

Baker (1988) proposed that not only solids holdup at 0° should be considered for predicting the design load. In the model developed by the author, masses at different angular positions in the upper half of the drum should also be considered. According to the hypothesis formulated by Glikin (1978), in a dryer operating with ideal load any flight in the lower half of the drum is the mirror image of that positioned vertically above it in the upper half, giving rise to Equation (2.19):

$$H_{\text{TOT}} = \left(2 \sum_{i=0}^{\text{LUF}} h_i \right) - h_{\text{FUF}} \qquad (2.19)$$

where i varies from 0° to the last discharge position of the flight.

It is worth mentioning that Baker's model (1988) can calculate only the mass of solids in the dense phase. The author assumed that the mass of solids in the airborne phase is approximately 10% of that in the dense phase. Therefore, to predict the design load, the results were multiplied by 1.1.

Ajayi and Sheehan (2012) proposed a modification to Baker's model (1988). The authors found that the predicted design load using models from the literature was different from those experimentally observed under several experimental conditions. For such reason, a correction factor of 1.24 was applied to the original model. In addition, in order to determine the total solids holdup in the design load condition, the ratio of the mass of solids in the active phase to the mass of solids in the passive phase (Y) was estimated through the average fall time of particles. The model developed by the authors is presented in Equation (2.20):

$$H_{\text{TOT}} = \left(1.24 \left(2 \sum_{i=0}^{\text{LUF}} h_i \right) - h_{\text{FUF}} \right) (1 + Y) \qquad (2.20)$$

Ajayi and Sheehan (2012) used two-segmented flights with an angle of 124° between the segments in their experiments. Likewise, Karali et al. (2015) also used the same type of flights, but with an angle between segments to create a more recent model. Using their experimental data, the authors made an adjustment similar to that proposed by Ajayi and Sheehan (2012). However, the Y factor was neglected in this model, because according to them in addition to being a difficult measurement to be performed, involving many uncertainties, it presents very low values (0.042–0.078), which would change the result very little. The model fitted to the data by Karali et al. (2015) is given by Equation (2.21):

$$H_{\text{TOT}} = \left(1.38 \left(2 \sum_{i=0}^{\text{LUF}} h_i \right) - h_{\text{FUF}} \right) \tag{2.21}$$

Nevertheless, the authors observed an ideal holdup of 5% when working with flights with a segment ratio of 0.375. This holdup value is outside the ideal range recommended by Perry and Green (1997), which is between 10% and 15%. Still, according to Perry and Green (1997), the peripheral speed of the drum must be between 0.25 and 0.5 m/s, which is different from that used by Karali et al. (2015) (from 0.07 to 0.12 m/s).

Nascimento et al. (2018) used peripheral speeds closer to the recommended range, that is, from 0.12 to 0.20 m/s. For these rotational speeds, the model presented by Equation (2.22) was adjusted as follows:

$$H_{\text{TOT}} = \left(1.59 \left(2 \sum_{i=0}^{\text{LUF}} h_i \right) - h_{\text{FUF}} \right) \tag{2.22}$$

The authors reported an adjusted parameter of 1.59, which is greater than the value of 1.38 suggested by Karali et al. (2015). Although both models were adjusted for the same type of flights, higher peripheral speeds tend to increase the quantity of solids in the active region of the drum, consequently leading to increased values of this parameter as a function of the rotational speed. Table 2.1 shows the model deviations from the experimental design load determined by Nascimento et al. (2018).

Apart from the rotational speed, particle properties and flight configurations can change ideal load conditions of the equipment. Nascimento et al. (2018) varied the drum rotational speeds so that the ideal load of the equipment ranged between 10% and 15%. These results will be explored in the following section.

2.3.4   ROTATIONAL SPEED FOR THE DETERMINATION OF THE DESIGN LOAD

According to Mujumdar (2014), it has already been empirically proven that the amount of solids inside rotary dryers must occupy between 10% and 15% of their volume so that the performance of the equipment is optimized. Based on this, all flights and operational variables must be designed in order to achieve the best load distribution and consequently improve drying rates.

Nascimento et al. (2018) analyzed the rotational speed at which the drum operated under the design load condition and found that the solids holdup occupied between

TABLE 2.1
Geometric Models to Estimate the Design Load: Model Deviations from Experimental Data.

Material	Rotational Speed (rpm)	Experimental Design Load (%)	Model Deviations from Experimental Data				
			Porter (1963) (%)	Kelly and O'Donnel (1977) (%)	Ajayi and Sheehan (2012) (%)	Karali et al. (2015) (%)	Nascimento et al. (2018) (%)
Filtered sand	22.8	10.5	−31.6	−25.9	−13.3	−3.6	10.5
Glass beads	23.5	10.5	−37.3	−32.1	−21.1	−12.2	1.0
Granulated sugar	12.2	10.5	−20.3	−13.7	−6.1	4.5	20.0
Filtered sand	43.8	13	−43.0	−38.3	−24.4	−15.9	−3.1
Glass beads	38.7	13	−45.5	−41.0	−33.2	−25.6	−10.0
Granulated sugar	34.8	13	−33.3	−27.7	−11.1	−1.1	12.3
Filtered sand	64.2	15.5	−51.6	−47.6	−32.4	−24.8	−13.5
Glass beads	61.4	15.5	−48.0	−43.7	−30.4	−22.6	−11.0
Granulated sugar	54.5	15.5	−41.0	−36.1	−18.7	−9.6	1.9

10% and 15% of the drum volume. The authors observed Froude numbers from 0.01 to 0.3, which made the equipment with a diameter of 0.108 m reach peripheral speeds between 0.08 and 0.40 m/s, falling within the range recommended by Perry and Green (1997).

From the experimental data, an empirical model was fitted to determine the ideal Froude number as a function of the drum filling degree and the physical and flow properties of the materials. The parameters associated with the physical properties were adjusted to find a representative model, as presented in Equation (2.23):

$$Fr_{\text{ideal}} = \frac{15}{\mu_{\text{din}}} f^2 - 6.5 \times 10^4 \frac{1}{\rho_s \sigma} f - 0.103 \tag{2.23}$$

where Fr_{ideal} is the ideal Froude number, μ_{din} is the coefficient of dynamic friction, ρ_s is the material density, σ is the static angle of repose, and f is the drum filling degree.

A comparison between the experimental data and the model in Equation (2.23) is illustrated in Figure 2.15.

By analyzing the parameters of Equation (2.7), it can be concluded that the ideal Froude number is lower for materials with a higher coefficient of dynamic friction due to the need for lower rotational speeds (centripetal force) to balance with the friction forces and keep the particles in the flights. Additionally, density can be considered a significant parameter. The correlation coefficient of the data was 0.92, pointing to a good agreement with the experimental results, with a maximum deviation of 4.4%.

Similarly, other models were fitted for 15 and 18 flights, as shown in Equations (2.24) and (2.25), respectively:

$$Fr_{\text{ideal}} = \frac{6.6}{\mu_{\text{din}}} f^2 - 3.2 \times 10^4 \frac{1}{\rho_s \sigma} f - 0.065 \tag{2.24}$$

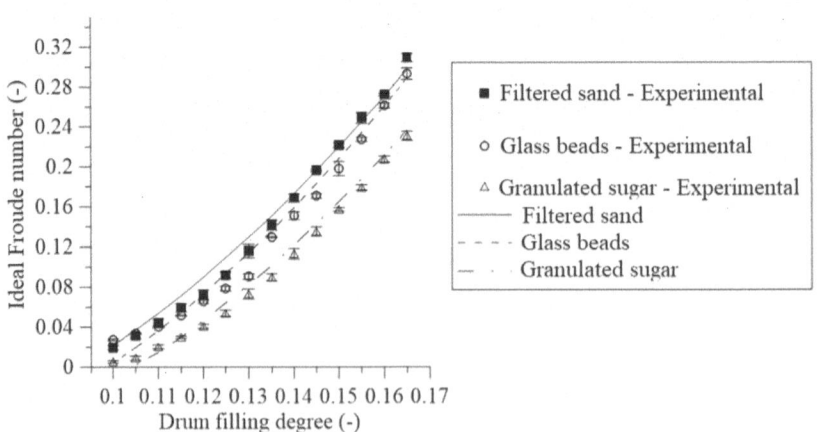

FIGURE 2.15 Comparison between experimental results and the model presented in Equation (2.23) (Nascimento et al., 2018).

$$Fr_{\text{ideal}} = \frac{3.6}{\mu_{\text{din}}} f^2 - 1.9 \times 10^4 \frac{1}{\rho_s \sigma} f - 0.046 \qquad (2.25)$$

A comparison between the experimental results and those obtained by Equations (2.24) and (2.25) is presented in Figure 2.16.

The model shows a good correlation coefficient for 15 flights ($r_2 = 0.92$), while for 18 flights the results were not well fitted. According to the authors, this can be attributed to the theoretical number of flights, which should not be greater than 16 at various rotational speeds analyzed for these particles. The definition of the theoretical number of flights will be covered in the next topic.

A model that includes the number of flights was then fitted, resulting in Equation (2.26):

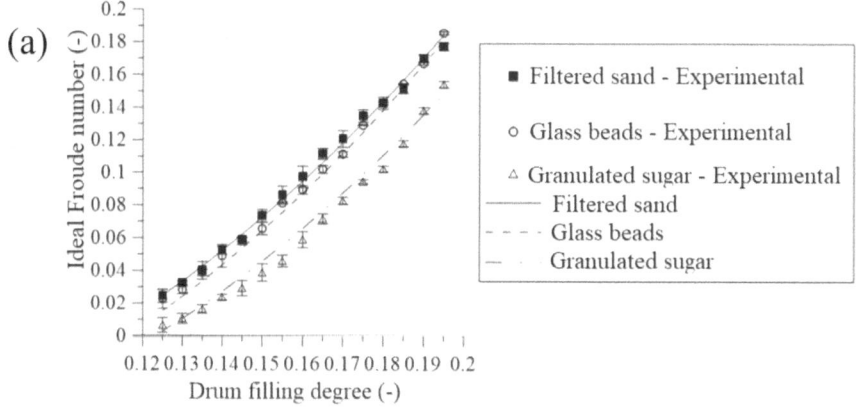

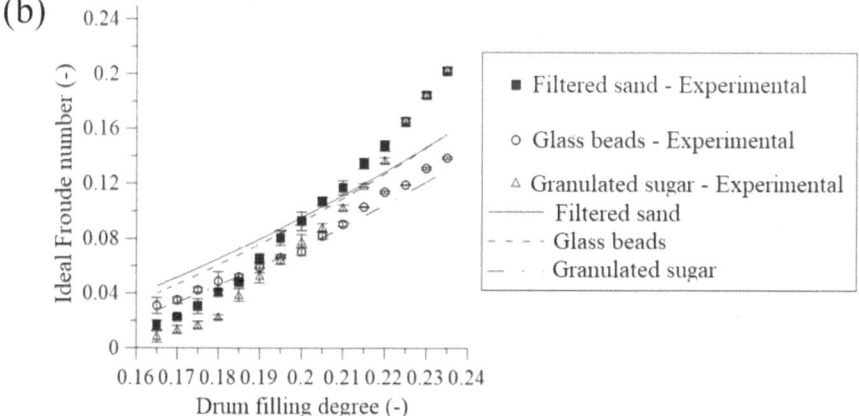

FIGURE 2.16 Comparison between the experimental results of ideal Froude number as a function of drum filling degree for dryers with (a) 12 flights and (b) 15 flights.

$$Fr_{ideal} = \frac{9.2 \times 10^4}{\mu_{din} n^{3.5}} f^2 - 1.1 \times 10^8 \frac{1}{\rho \sigma n^3} f - \frac{16}{n^2} \qquad (2.26)$$

where n is the number of flights in the equipment.

The correlation coefficient of this fit was 0.94. This is a relevant correlation since through Equation (2.26) it is possible to calculate the Froude number at which a rotary drum must operate for a given material as well as the number of flights in order to have a better use of the useful drying area.

2.3.5 THEORETICAL NUMBER OF FLIGHTS

The theoretical number of flights is the maximum number of flights that must be contained in a rotary drum so that an adjacent flight does not interfere with the load of the other and carry as many particles as possible.

Sunkara et al. (2013), defined an equation to predict the theoretical number of two-segmented flights and the right angle between segments. The parameters defined by the authors are illustrated in Figure 2.17.

In Figure 2.17, α is the angle between the lines that pass through both ends of the flight and the center of the drum and is calculated by Equation (2.27):

$$\tan \alpha = \frac{l_2}{r_h} \qquad (2.27)$$

The angle υ is the one formed between the highest point occupied by the particles in the flight and the line that divides the drum in half, being established as a function of

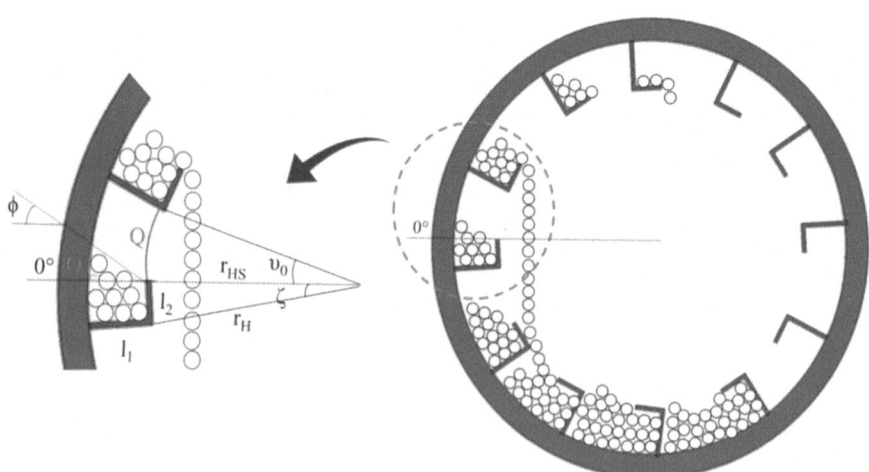

FIGURE 2.17 Definition of important parameters to determine the theoretical number of flights (Silveira et al., 2020).

the dynamic angle of repose at the 0° position $\left(\phi\,|_{\delta_L=0°}\right)$ and calculated by Equation (2.28):

$$\tan \upsilon = \left(1 - \frac{\dfrac{r_H}{R}}{\cos \alpha}\right) \tan \phi\,|_{\delta_L=0°}$$

(2.28)

With the calculated values of the parameters α and υ it is possible to estimate the minimum spacing that must exist between flights and consequently the maximum number of flights, which is a theoretical number, using Equation (2.29):

$$n_{\text{design}} = \frac{360°}{\alpha + \upsilon}$$

(2.29)

2.3.6 LAST UNLOADING FLIGHT

The Last Unloading Flight (LUF) is the angle where the curtain of particles ends in a dryer, as shown in Figure 2.18. LUF appears as a complementary tool to establish the ideal load, as it analyzes where the discharge of solids starts. The greater the angle of last discharge, the more distributed the solids over the active phase of the drum.

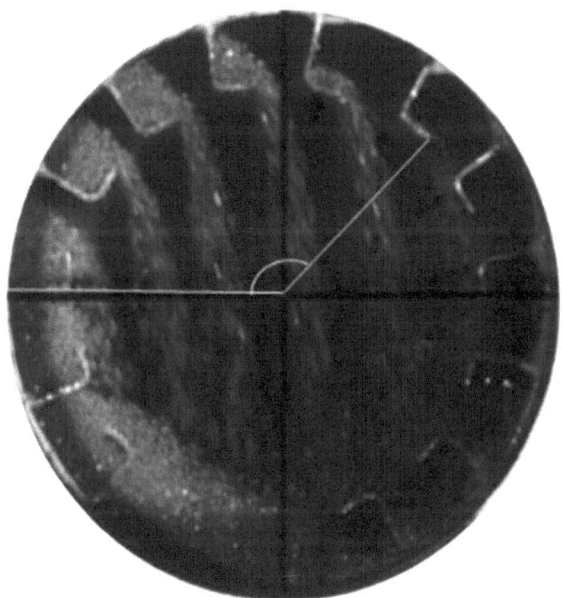

FIGURE 2.18 Measurement of the angular position of the last unloading flight.

The last unloading flight increases with increasing the rotational speed, that is, higher rotational speeds lead to an increment in solids holdup in the flights, which slows down their discharge movement.

In addition, materials with higher static angles of repose have greater resistance to flow and discharge, also causing an increase in the last unloading flight.

Nascimento et al. (2018) analyzed the last unloading flight of two-segmented flights as a function of the Froude number for glass beads, filtered sand, and granulated sugar. The authors observed that the prediction of the last unloading flight could be made by a logarithmic model, according to Equation (2.29):

$$LUF\ (^\circ) = 206.9\ \frac{1}{\gamma}\ln Fr + 169.7 \qquad (2.29)$$

where γ is the coefficient of dynamic friction of the material.

A comparison between the results obtained by Equation (2.29) and the experimental data is shown in Figure 2.19.

The LUF position increases fast with increasing the Froude number for lower rotational speeds, while for higher speeds such increase is smaller, suggesting a logarithmic dependence between the rotational speed for the design load and the drum filling degree.

When the drum is loaded below the design load, the curve of the mass of solids in the flight as a function of the angular position remains constant until the flight saturates and this curve intersects that of the ideal load, as observed in Figure 2.20.

Therefore, regardless of the number of flights and the load of solids in the drum, since the flight will always carry as much mass of solids as possible for that angular

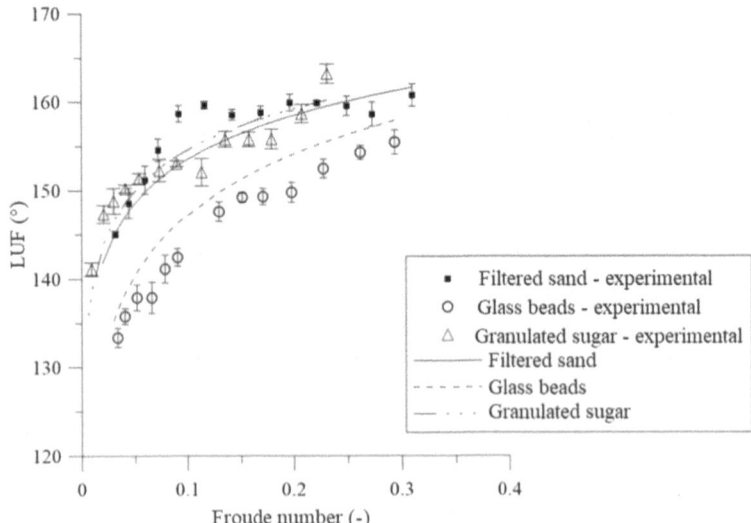

FIGURE 2.19 Comparison between the last unloading flight obtained experimentally and through Equation (2.29).

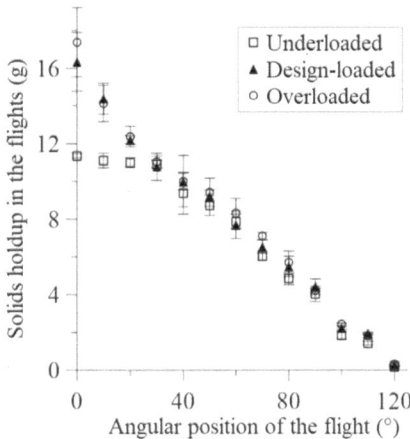

FIGURE 2.20 Mass of solids in the flights for a drum operating under different loading conditions at the same rotational speed.

position, the discharge of solids is dependent only on the rotational speed and the material inside the equipment.

REFERENCES

Ajayi, O.O. Multiscale modelling of industrial flighted rotary dryers, Ph.D. thesis, James Cook University (2011), School of Engineering, Australia.

Ajayi, O.O.; Sheehan, M.E. "Design loading of free flowing and cohesive solids in flighted rotary dryers" *Chemical Engineering Science* 73 (2012): 400–411. https://doi.org/10.1016/j.ces.2012.01.033.

Baker, C.G.J. "The design of flights in cascading rotary dryers" *Drying Technology* 6 (1988): 631–653. https://doi.org/10.1080/07373938808916402.

Glikin, P.G. "Transport of solids through flighted rotation drums" *Trans IChemE* 56 (1978): 120–126.

Karali, M.A. Analysis study of the axial transport and heat transfer of a flighted rotary drum operated at optimum loading, Ph. D. thesis, Otto Von Guericke University (2015), Magdeburg, Germany.

Kelly, J.J. Bull. Inst. Ind. Res. Standards 5 (1968): 361 p.

Kelly, J.J.; O'Donnell, J.P. "Residence time model for rotary drums" *Trans IChemE*, 55 (1977): 243–252.

Konidis, J. Design of Direct Heated Rotary Dryers, A Major Technical Report, The Department of Mechanic Engineering. Concordia University (1984), Montreal/Quebec.

Lisboa, M.H.; Vitorino, D.S.; Delaiba, W.B.; Finzer, J.R.D.; Barrozo, M.A.S. "A study of particle motion in rotary dryer" *Brazilian Journal of Chemical Engineering* 94 (2007): 365–374. https://doi.org/10.1590/S0104-66322007000300006.

Matchett, A.J., Baker, C.G.J. "Particle residence times in Cascading rotary dryers. Part 2: Application of the two-stream model to experimental and industrial data" *Journal of Separation Process Technology* 9 (1988): 5.

Matchett, A.J., Sheikh, M.S. "An improved model of particle motion in cascading rotary dryers" *Trans IChemE* 68 (1990): 139.

Mujumdar, A.S. *Handbook of Industrial Drying*, CRC Press, Boca Raton, FL, 2014. https://doi.org/10.1201/b17208.

Mujumdar, A.S.; Krokida, M.; Marinos-Kouris, D., Rotary drying. In: *Handbook of Industrial Drying*, CRC Press, Boca Raton, FL, 3rd ed., 2007.

Nascimento, S.M.; Duarte, C.R.; Barrozo, M.A.S. "Analysis of the design loading in a flighted rotating drum using high rotational speeds" *Drying Technology* 36 (2018): 1200–1208. https://doi.org/10.1080/07373937.2017.1392972.

Perry, R.H.; Green, D.W. *Perry's Chemical Engineers' Handbook*, 7th ed., Mc-Graw-Hill, New York, 1997.

Porter, S.J., "The design of rotary dryers and coolers" *Transport Institute of Chemical Engineering* 41 (1963): 272–287.

Revol, D.; Briens, C.L., Chabagno, J.M., "The design of flights in rotary dryers" *Powder Technology* 121 (2001): 230–238. https://doi.org/10.1016/S0032-5910(01)00362-X.

Schofield, F.R.; Glikin, P.G. "Rotary coolers for granular fertilizer" *Chemical and Process Engineering Resources* 40 (1962): 183.

Silveira, J.C.; Brandão, R.J.; Lima, R.M.; Machado, M.V.C.; Barrozo, M.A.S.; Duarte, C.R. "A fluid dynamic study of the active phase behavior in a rotary drum with flights of two and three segments" *Powder Technology* 368 (2020): 297–307. https://doi.org/10.1016/j.powtec.2020.04.051.

Silveira, J.C.; Brandão, R.J.; Lima, R.M.; Duarte, C.R.; Barrozo, M.A.S. "A study of the design and arrangement of flights in a rotary drum" *Powder Technology* 395 (2022): 195–206. https://doi.org/10.1016/j.powtec.2021.09.043.

Silvério, B.C.; Arruda, E.B.; Duarte, C.R.; Barrozo, M.A.S. "A novel rotary dryer for drying fertilizer: Comparison of performance with conventional configurations" *Powder Technology*, 270 (2015): 135–140. https://doi.org/10.1016/j.powtec.2014.10.030.

Sunkara, K.R.; Herz, F.; Specht, E.; Mellmann, J. "Influence of flight design on the particle distribution of a flighted rotating drum" *Chemical Engineering Science*, 90 (2013): 101–109. https://doi.org/10.1016/j.ces.2012.12.035.

3 Nonconventional Rotary Dryers

3.1 INTRODUCTION

With the increasing focus on the reduction of greenhouse gas emissions and energy demand, the design of drying units has become critical. Despite their simplicity and flexibility, conventional rotary dryers represent a significant cost to industry (Sheehan et al., 2005). The scale-up of rotary dryers from pilot scale and the product quality control during processing disturbances remain complex and poorly described (Baker, 1988). Because of the complexity in estimating some parameters, most rotary dryers are overdesigned, making the final product overdried and overheated, and consequently wasting thermal energy and increasing costs. Due to these difficulties, conventional rotary dryers have undergone some modifications in order to improve their performance (Arruda, 2008).

The performance of conventional flighted rotary dryers depends primarily on the effectiveness of contact between the cascading particles and the drying gas across the drum length. The cascade of solids from the flights is called active or dilute phase and takes place in the active region of the dryer where the fluid–particle contact is enhanced (Britton et al., 2006). The solids resting in the flights and on the drum floor are referred to as passive or dense phase and occur in the passive region of the device where the fluid–particle contact is poor (Kemp, 2004).

High operational efficiency can be achieved with optimal loading of the drum, increased solids residence time, suitable mixture of the bed (ensuring uniform temperature and moisture of the material), and maximum utilization of the active region where the drying occurs in greater magnitude (Silveira et al., 2022). Modifications in the drying configurations to promote an appropriate cascade pattern of particles (Silvério et al., 2015), a high amount of solids in the active region (Silveira et al., 2020), and a homogeneous material distribution over the cross-section of the dryer (Nascimento et al., 2019) are highly desirable. Furthermore, deeper modifications to enhance the time of effective fluid–particle contact have also been recently investigated (Arruda et al., 2009a).

Other studies in the literature have considered adaptations in the rotary drum design for specific applications. For example, for the drying of fine and brittle particles using a flighted rotary drum (FRD) with a large cross-section, internal elements or partitions have been used to increase the material distribution and reduce dusting and impact grinding (Silvério, 2012). In contrast, for granular materials with low flowability the spiral flights have been positioned at the beginning (near the feed) and end (near the discharge) of the drum to accelerate the material flow (Keey and Danckwerts, 2013).

DOI: 10.1201/9781003258087-3

The particulate system research group of the Federal University of Uberlandia developed another version of the rotary dryer, named roto-aerated dryer (Lisboa et al., 2007). The main feature of this new dryer is the presence of an aerated system consisting of a central pipe (encased in the drum) from which a series of mini pipes take hot air directly to the bed of particles flowing at the bottom (without flights). This reduces the residence time of particles, with a subsequent increase in the processing capacity (Fernandes et al., 2009).

Another important point that should be considered is that the use of conventional rotary dryers is limited to granular materials with appropriate flowability and may not be used to process pasty materials. However, recent modifications in conventional rotary dryers (Silva et al., 2019b) have been made to allow the use of this type of dryer in a nonconventional configuration for suspension drying.

In this monograph, two main configurations of nonconventional rotary dryers are highlighted, one intended to increase the effective time of the fluid–particle contact, i.e., the roto-aerated dryer, and another designed for possible drying of pastes, i.e., the rotary dryer with inert filling.

3.2 NONCONVENTIONAL CONFIGURATIONS: DESIGN AND APPLICATIONS

3.2.1 ROTO-AERATED DRYERS

To improve the effective contact time between hot air and wet solids and consequently the drying efficacy, another version of the rotary dryer, known as "roto-fluidized" dryer, or more accurately "roto-aerated" dryer was evaluated by Lisboa et al. (2007). As previously mentioned, the major characteristic of this dryer configuration is its aerated system consisting of a central pipe (surrounded by the drum) from which a series of mini pipes conduct the hot air directly to the bed of particles flowing at the bottom of the surrounding drum (without flights). Figure 3.1 shows a schematic diagram of air distribution in this new dryer.

The gas–particle contact in the roto-aerated dryer is highly effective and occurs as long as the solids remain in the dryer, unlike the conventional cascading rotary dryer, in which this contact occurs mostly when the particles are falling from the flights (active region). This fact causes a reduction in the particle residence time with

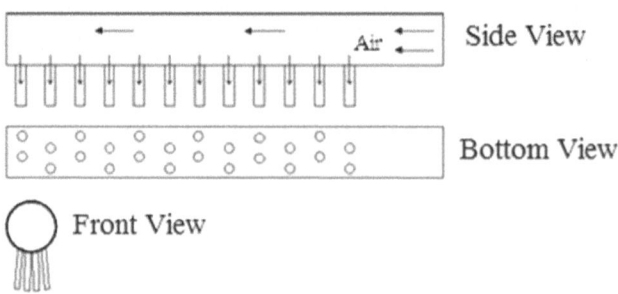

FIGURE 3.1 Schematic diagram of air distribution in the roto-aerated dryer.

a subsequent increase in the processing capacity, as observed by Lisboa et al. (2007). Moreover, greater transfer coefficients of mass and energy were observed in preliminary studies (Lisboa et al., 2007), resulting in enhanced efficacy of the roto-aerated dryer in relation to the conventional equipment.

a. Drying of fertilizers:

Arruda et al. (2009) compared the performance of a conventional rotary dryer to that of a roto-aerated dryer. The experimental trials performed in the conventional dryer used two different configurations: one with two-segmented flights (Conv. F. 2 seg) with dimensions L1 = 0.03 m and L2 = 0.01 m and the other with three-segmented flights (Conv. F. 3 seg) with dimensions L1 = 0.02 m and L2 = L3 = 0.007 m. The flights had a length of 1.5 m and an angle between segments of 135o. The experiments carried out with the roto-aerated dryer were also based on different configurations: one with 56 mini pipes with an internal diameter of 9 mm (RF – 9 mm) and the other with 56 mini pipes with a diameter of 20 mm (RF – 20 mm). The particles in this study (Arruda et al., 2009b) were fertilizers in the form of simple super-phosphate granules (SSPG) with a Sauter mean diameter of 2.45 mm, a particle density of 1,100 kg m^{-3}, and a heat capacity of 0.245 kcal kg^{-1}°C, and initial moisture contents between 0.12 and 0.15 kg water/kg dry solid.

The results of the residence time for the comparison performed by Arruda et al. (2009) are plotted in Figure 3.2 and the conditions of these experiments are shown in Table 3.1. It can be observed that the residence times for the roto-aerated dryer were about 48% smaller than for the conventional dryer.

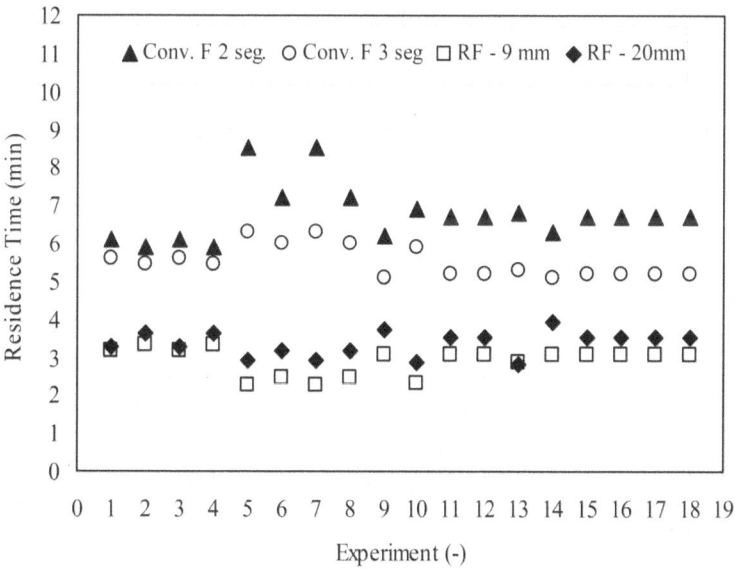

FIGURE 3.2 Residence time results for each configuration of roto-aerated and conventional rotary dryers.

TABLE 3.1

Operating Conditions of Each Experiment Carried Out by Arruda et al. (2009)

Experiment	v_f (m s^{-1})	T_f (°C)	G_s (kg min^{-1})
1	1.50	75.0	0.8
2	1.50	75.0	1.2
3	1.50	95.0	0.8
4	1.50	95.0	1.2
5	3.50	75.0	0.8
6	3.50	75.0	1.2
7	3.50	95.0	0.8
8	3.50	95.0	1.2
9	1.09	85.0	1.0
10	3.91	85.0	1.0
11	2.50	70.9	1.0
12	2.50	99.1	1.0
13	2.50	85.0	0.72
14	2.50	85.0	1.28
15	2.50	85.0	1.0
16	2.50	85.0	1.0
17	2.50	85.0	1.0
18	2.50	85.0	1.0

Hence, the processing capacity of this new equipment is also expected to improve when compared with the conventional cascading rotary dryer. In the conventional rotary dryer, the fact that particles are lifted in the flights, slide, roll, and then fall in spreading cascades through an air stream and reenter the bed at the bottom, possibly bouncing and rolling (Silvério, 2012), determines the levels of residence time. In contrast, the particle motion in the roto-aerated dryer is simpler, implying a lower residence time.

Figure 3.3 shows the drying rates obtained in the conventional and roto-aerated configurations studied by Arruda et al. (2009). As expected, the drying rate of the conventional rotary dryer was higher for three-segmented flights since this type of configuration tends to promote a more homogeneous cascade of solids, thus reducing the presence of dead zones inside the dryer.

The results of Figure 3.3 also show that the best configuration for the roto-aerated dryer was that with mini pipes with a diameter of 9 mm (RF – 9 mm). As observed, a reduction in the mini pipes diameter promotes higher speeds of hot air and therefore larger drying rates. It can also be seen that the drying rate values of the roto-aerated dryer were between 3.1 and 4.9 times higher than those of the conventional cascading rotary dryer.

Figure 3.4 presents the solids temperature difference, i.e., the difference between the outlet and inlet temperatures of solids for each dryer

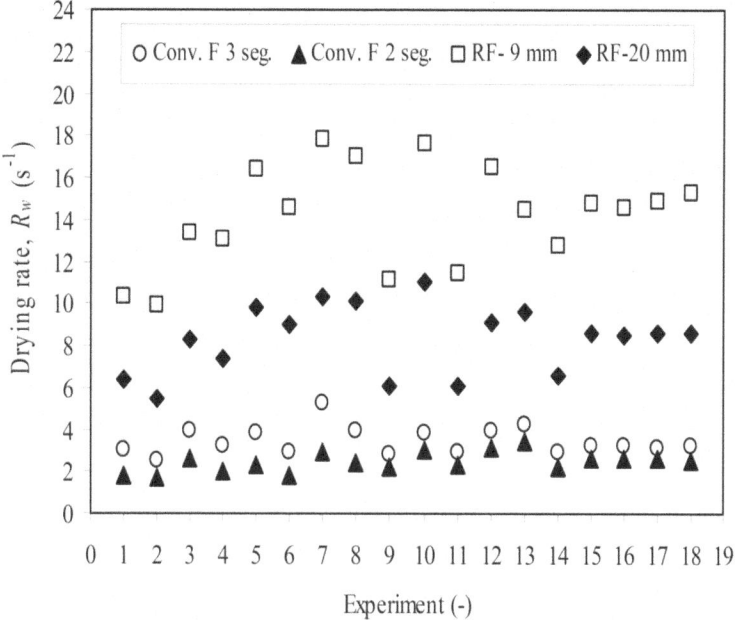

FIGURE 3.3 Drying rate results for each configuration of roto-aerated and conventional rotary dryers (conditions in Table 3.1).

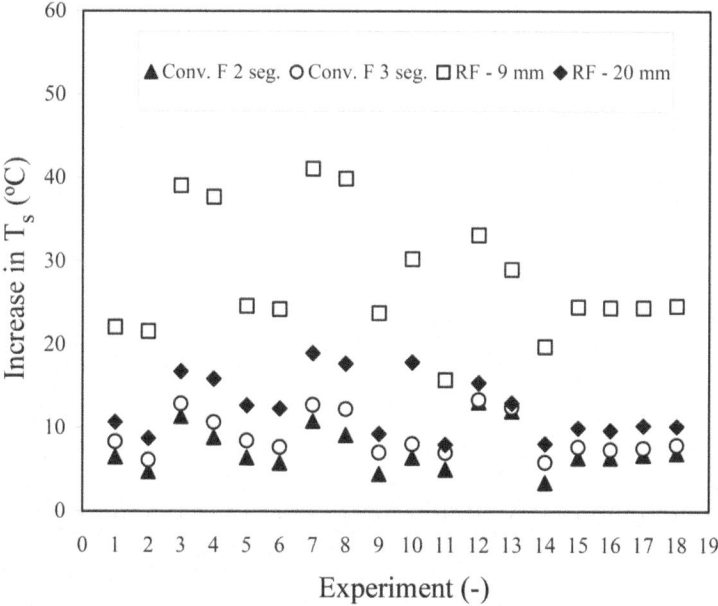

FIGURE 3.4 Difference between the outlet and inlet temperatures of solids for each configuration of roto-aerated and conventional rotary dryers (conditions in Table 3.1).

configuration. For the roto-aerated dryer, the solids temperature differences were 1.7–3.3 times higher than those for the conventional cascading rotary dryer, indicating an enhanced heat transfer efficiency. However, it must be taken into account that higher temperatures can be harmful for thermosensitive materials.

Silvério et al. (2011) changed the configuration of the 56 mini pipes in a roto-aerated dryer, named hybrid roto-aerated dryer or RT-96, resulting in the following distribution: 28 mini pipes with a diameter of 0.009 m and 28 mini pipes with a diameter of 0.006 m. The 56 mini pipes were arranged in an interleaved configuration, as shown in Figure 3.5. Table 3.2 lists the operating conditions used in the experiments conducted by Silvério et al. (2011) for fertilizer drying.

Figure 3.6 displays the residence time in the hybrid roto-aerated and conventional rotary dryers (concurrent and countercurrent flows) obtained by Silvério et al. (2011) for each operating condition (Table 3.2).

The residence time in the hybrid roto-aerated dryer (RT-96) was lower than that obtained in the conventional dryers. The conventional rotary dryers have four components of particle movement: (a) gravitational due to drum inclination; (b) gas drag (negative in countercurrent flow and positive in concurrent flow); (c) bouncing of particles upon impact with the dryer bottom; and (d) rolling of particles at the dryer bottom, especially in overloaded dryers (Baker, 1988). On the other hand, the particle motion in the roto-aerated dryer is simpler since it is influenced mainly by gas drag and gravitational forces (due to drum inclination). These results indicate that the difference between the newly developed dryer (RT-96) and the conventional one raises with the increase in the airflow rate due to the consequent increment in drag force. The greatest differences were found in experiments 5, 6, 7, 8, and 10, which were carried out at higher airflow rates. In these conditions, the residence times in the RT-96 were up to 10-fold lower than those obtained in the conventional dryers. Therefore, the processing capacity of this new equipment is also expected to be superior to that of conventional cascading rotary dryers.

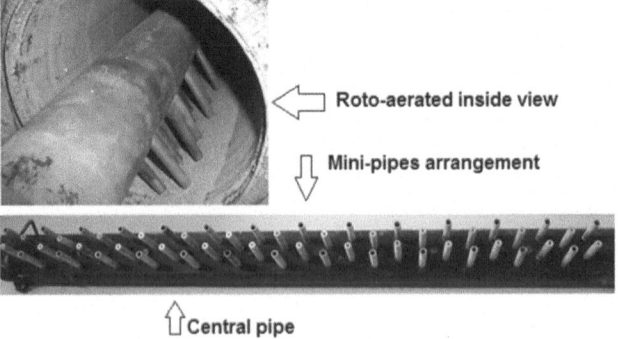

FIGURE 3.5   Schematic diagram of air distribution in the RT-96 hybrid roto-aerated dryer.

TABLE 3.2

Operating Conditions of Each Experiment Conducted by Silvério et al. (2011)

Experiment	G_f (m³·s⁻¹)	T_f (°C)	G_s (kg seg⁻¹)
1	0.05	75.0	0.013
2	0.05	75.0	0.020
3	0.05	95.0	0.013
4	0.05	95.0	0.020
5	0.11	75.0	0.013
6	0.11	75.0	0.020
7	0.11	95.0	0.013
8	0.11	95.0	0.020
9	0.03	85.0	0.017
10	0.12	85.0	0.017
11	0.08	70.9	0.017
12	0.08	99.1	0.017
13	0.08	85.0	0.012
14	0.08	85.0	0.022
15	0.08	85.0	0.017
16	0.08	85.0	0.017
17	0.08	85.0	0.017
18	0.08	85.0	0.017

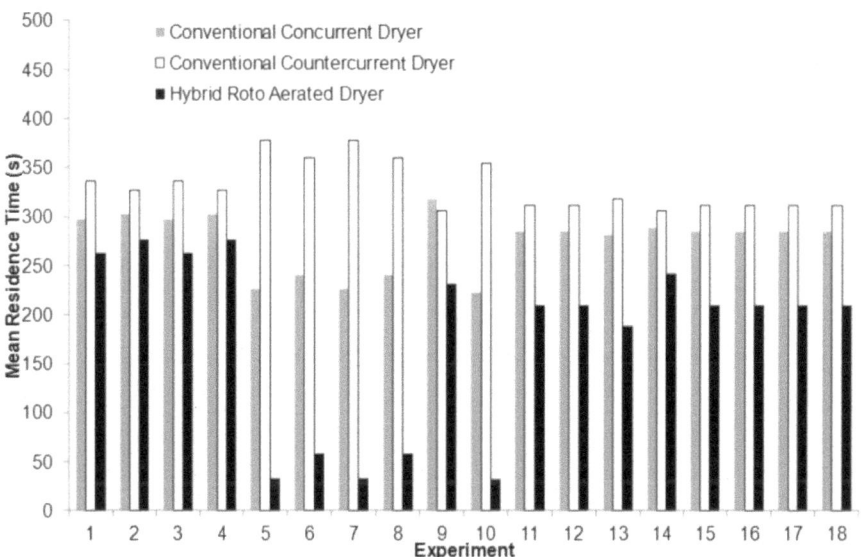

FIGURE 3.6 Residence time in the RT-96 and conventional rotary dryers (2011).

Figure 3.7 illustrates the drying rates obtained with the conventional and hybrid roto-aerated rotary dryers in all experiments conducted by Silvério et al. (2011). As it can be seen, the new mini pipes arrangement promotes higher hot air velocities than the previous one (Arruda et al., 2009b), leading to increased water removal rates, lower residence time, and consequently higher drying rates. Note that the drying rates in the experiments performed with the RT-96 dryer were much higher than those obtained with the conventional rotary dryers. This increase is even more pronounced in experiments 5, 7, and 10, which were performed at higher airflow rates. The drying rates in these operating conditions were respectively 15-, 16-, and 18-fold higher than those found in the conventional rotary dryers, evidencing the superior performance of the new dryer (RT-96).

Silvério et al. (2011) observed that the difference between the inlet and outlet temperatures of solids was 2–5-fold times higher in the roto-aerated dryer than in the conventional rotary dryers under the same operating conditions. This evidences the better gas–particle contact achieved in the roto-aerated dryer, resulting in higher heat transfer coefficients. However, these results also indicate that it is necessary to evaluate the experimental conditions when using this dryer to process heat-sensitive materials, e.g., some food products, in order to prevent product degradation.

b. Drying of fruit residues

The byproducts of fruit processing industry, such as seeds, kernels, and bagasse (which were previously considered wastes), have significant potential use as food supplementation (Duzzioni et al., 2013) as they can contain higher amounts of phenolic and other bioactive compounds than the

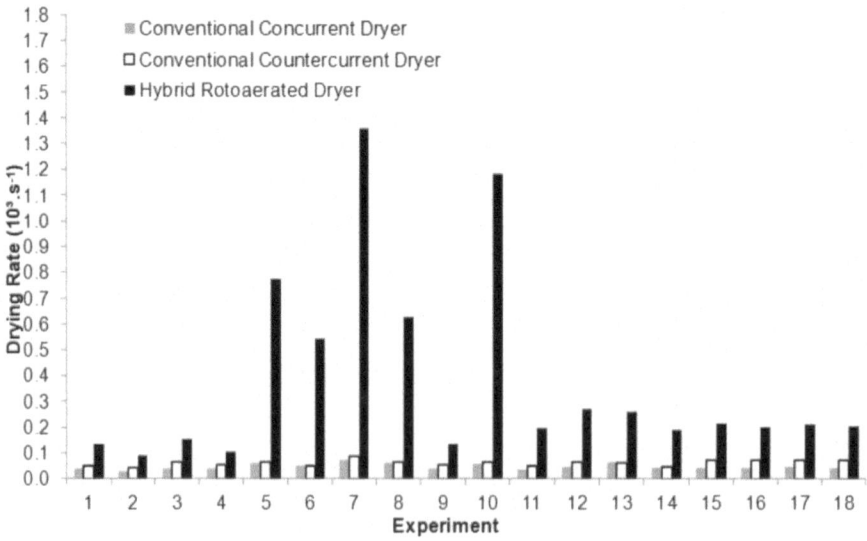

FIGURE 3.7 Drying rates obtained with each configuration of the RT-96 and the conventional rotary dryers (Silvério et al., 2011).

edible fleshy parts (Bortolotti et al., 2013). It is estimated that 40% of the volume of processed fruits are unused residues (Sousa et al., 2011). These aspects, along with the increasing global interest in environmental-friendly technologies, explain the recent effort to use byproducts of fruit processing industry (Silva et al., 2019a).

Fruit residues may contain more than 80% of water, which ends up limiting their shelf life and complicating their transport and storage (Silva et al., 2017). Thus, to reduce the moisture content it is necessary to submit these residues to a dehydration process. Nevertheless, since these materials are heat sensitive, it is essential to choose appropriate drying techniques and operating conditions in order to preserve the bioactive compounds present in the final product.

Silva et al. (2016) evaluated the viability of a roto-aerated dryer in the dehydration of acerola residues from the fruit processing industry, considering its potential use as food supplementation. The device was used in the RT-96 configuration (Silvério et al., 2011) and the effects of the process variables on the content of bioactive compounds were also investigated.

The results of the drying performance of acerola residues using this equipment (Silva et al., 2016) are presented in Figures 3.8 and 3.9 for each experiment of the experimental design (see conditions in Table 3.3).

The average residence time of acerola residues in the new dryer (roto-aerated) ranged from 3.2 to 4.4 min. It can be seen in Figure 3.8 that the drying rate (Rw) values were high in this new dryer. The best condition (highest drying rate and moisture removal) was the axial point for temperature

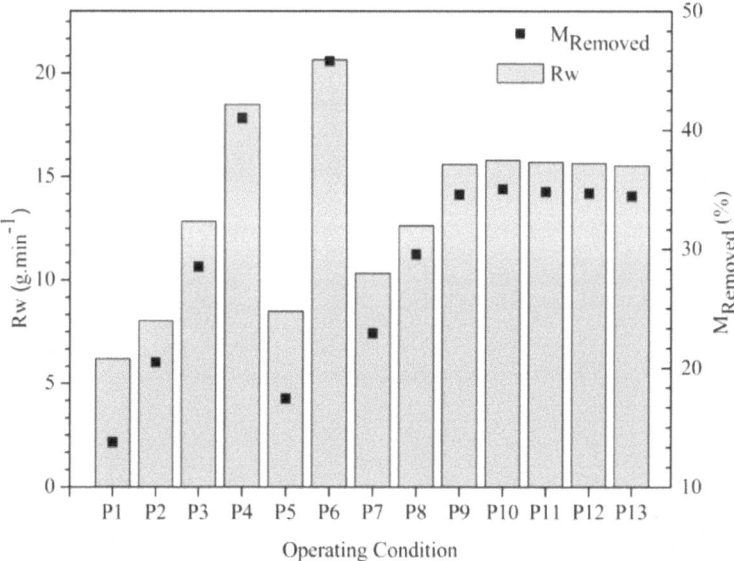

FIGURE 3.8 Drying rate (Rw) and moisture removal in each experiment of the acerola drying process (Table 3.3).

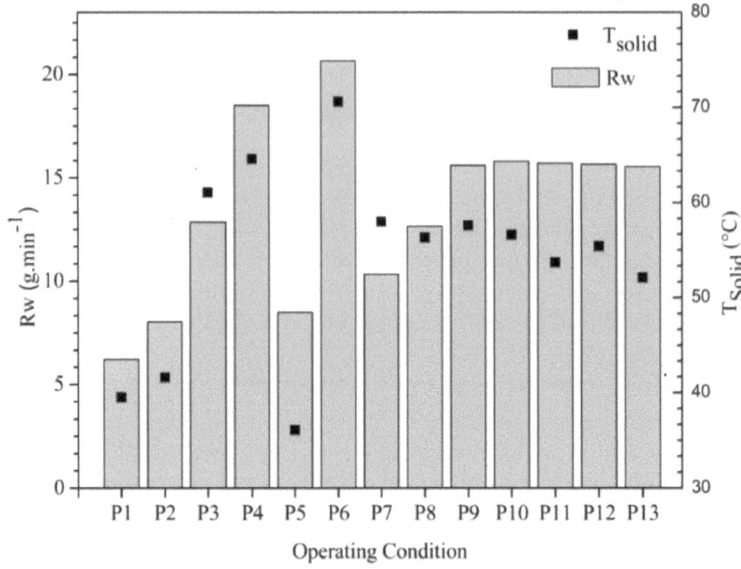

FIGURE 3.9 Drying rate (Rw) and solid temperature in each experiment of the acerola drying process (Table 3.3).

TABLE 3.3
Experimental Design of the Acerola Drying Process (Silva et al., 2016)

Experiment	T_f (°C)	v_f (m·s⁻¹)
P1	80.0	1.50
P2	80.0	3.00
P3	150.0	1.50
P4	150.0	3.00
P5	70.6	2.25
P6	159.3	2.25
P7	115.0	1.30
P8	115.0	3.20
P9	115.0	2.25
P10	115.0	2.25
P11	115.0	2.25
P12	115.0	2.25
P13	115.0	2.25

(P6 in Table 3.3). The drying rate and percentage of removed water in this condition were respectively more than 3 and 20 times higher than the experiment carried out at low levels of temperature and air velocity (P1). P6 was also the condition that led to the highest final temperature of the acerola residue (Figure 3.9). Thus, taking into account that the temperature of the material in a drying process can affect the final quality of the product (Silva et al., 2021; Nóbrega, 2012), it is also important to assess quality indices after drying.

The quality of acerola residues after drying using the RT-96 configuration of the roto-aerated dryer was compared to the values obtained with fresh samples by Silva et al. (2016), who used the quantification of total phenolic content (TPC) and total flavonoid content (TFC) to evaluate the samples. Figures 3.10 and 3.11 show the results of these bioactive compounds after drying for each operating condition of the experimental design proposed by Silva et al. (2016). The TPC and TFC values of the fresh sample were 1,266.8 ± 49.3 mg gallic acid per 100 g of dry residue and 4.68 ± 0.24 mg rutin per 100 g of dry residue, respectively.

According to Figure 3.10, the total phenolic content of acerola residues after drying (mean ± kf. SD) varied between 726.6 ± 14.2 mg gallic acid per 100 g and 913.6 ± 16.6 mg gallic acid per 100 g in Experiments P5 and P13, respectively, which correspond to 57.4% and 72.1% of the value obtained with the fresh sample (Silva et al., 2016). This result confirms the low degradation of TPC during drying in this novel device, mainly if performed under intermediate operating conditions (Chang et al., 2006).

Figure 3.11 shows that the highest total flavonoid content (4647.9 ± 165.3 μg rutin $100 g^{-1}$) was obtained in Experiment P8 (Table 3.1). This value was very close to that obtained with the fresh residue (Silva et al., 2006), indicating negligible degradation in this condition.

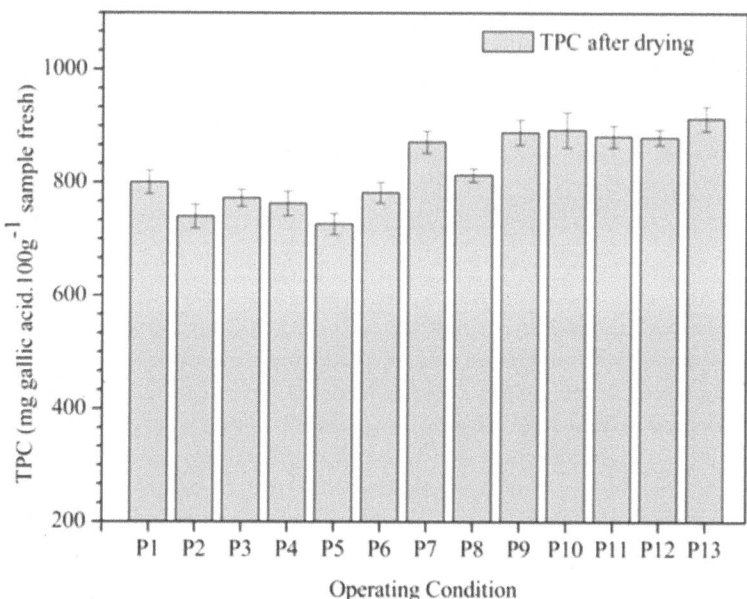

FIGURE 3.10 Total phenolic content of acerola residues before (fresh sample) and after drying under different operating conditions.

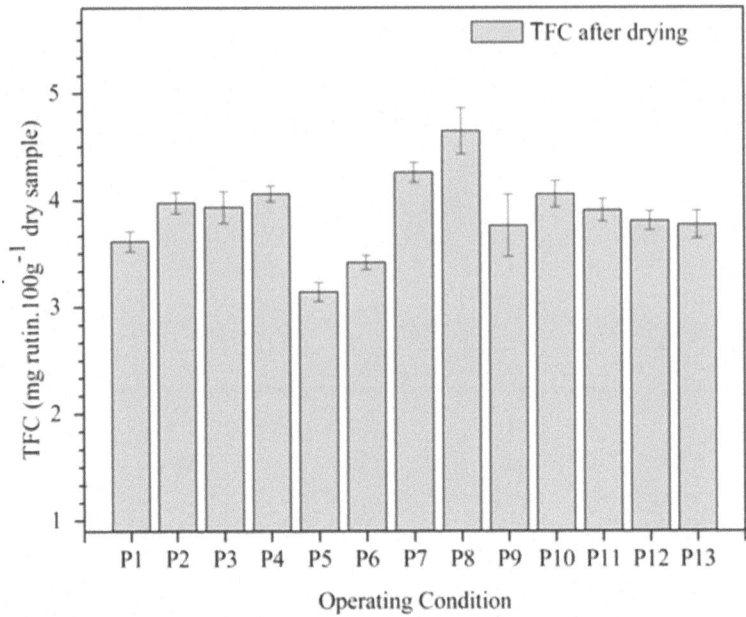

FIGURE 3.11 Total flavonoid content of acerola residues before (fresh sample) and after drying under different operating conditions.

Therefore, it can be concluded that this new dryer is able to efficiently dehydrate acerola residues. Additionally, the conditions that maximize the drying performance (Figures 3.8 and 3.9) are not the same that lead to the best quality parameters (Figures 3.10 and 3.11). However, it is possible to operate with this new dryer with an adequate drying performance, maintaining the high-quality levels of the final product.

c. Infrared roto-aerated dryer

Silva et al. (2021) investigated the drying of acerola residues in a roto-aerated dryer (RT-96 configuration) using a predrying system with infrared radiation combined with pretreatment with ethanol so as to perform the drying in this equipment in a single stage. The effect of operating variables (i.e., air velocity, air temperature, and infrared lamp power) on the drying performance and product quality in terms of content of bioactive compounds was evaluated by Silva et al. (2021).

Figure 3.12 illustrates the experimental apparatus of the roto-aerated dryer containing a predrying system with infrared radiation. The feed of solids was performed using a conveyor belt mounted below a reservoir, and a system with four infrared lamps (6) was mounted on a second conveyor belt in order to predry the acerola residues.

Table 3.4 shows the experimental results of the drying experiments using a roto-aerated dryer containing a predrying system with infrared radiation (Silva et al., 2021). The initial moisture content of acerola residues used in these experiments was 78.7 ± 0.8 g/100 g w.b. The fifth column of Table 3.4

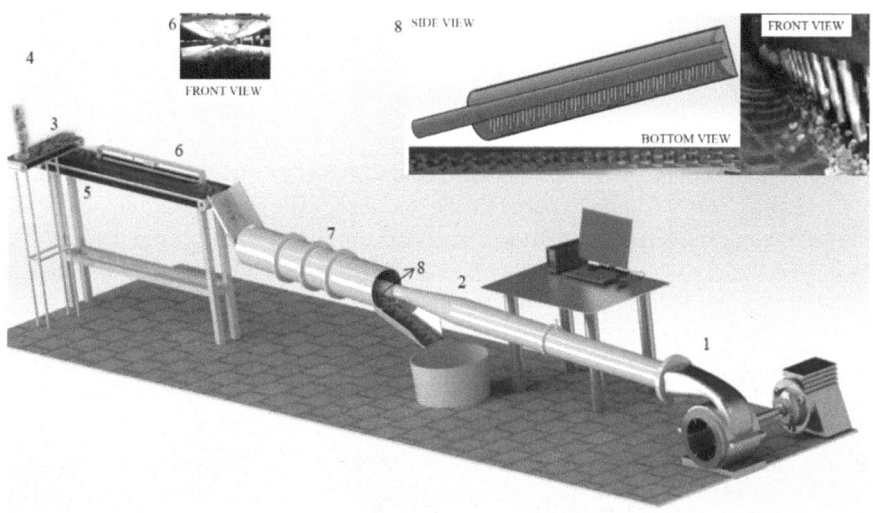

FIGURE 3.12 Experimental apparatus of the roto-aerated dryer containing a predrying system with infrared radiation (Silva et al., 2021).

TABLE 3.4
Levels and Results of the Central Composite Design

Exp.	T (°C)	v (m s⁻¹)	P (W)	M_{ir} (g/100 g)	M_{roto} (g/100 g)	Holdup (g)	R_w (g min⁻¹)	τ (g min⁻¹)
1	80.0	1.5	600.0	77.1	65.8	183.7	25.6	6.1
2	80.0	1.5	1100.0	64.0	49.0	97.0	25.3	6.0
3	80.0	3.0	600.0	77.0	63.9	194.2	24.1	6.8
4	80.0	3.0	1100.0	64.3	52.8	93.8	24.2	5.0
5	150.0	1.5	600.0	77.0	57.2	168.1	27.4	5.7
6	150.0	1.5	1100.0	64.2	42.8	98.3	28.1	5.2
7	150.0	3.0	600.0	78.0	46.6	122.0	30.0	6.2
8	150.0	3.0	1100.0	66.2	42.1	83.2	29.3	4.7
9	67.6	2.3	850.0	69.1	64.0	120.3	21.8	3.7
10	162.4	2.3	850.0	69.7	48.1	76.8	30.5	3.1
11	115.0	1.3	850.0	69.5	57.4	145.1	24.0	5.4
12	115.0	3.3	850.0	70.8	51.5	131.7	23.8	5.6
13	115.0	2.3	511.7	76.9	62.3	124.1	29.4	3.9
14	115.0	2.3	1188.3	59.1	36.6	68.8	31.3	4.3
15	115.0	2.3	850.0	68.9	55.7	73.2	25.2	3.1
16	115.0	2.3	850.0	70.9	55.5	70.7	26.2	3.2
17	115.0	2.3	850.0	68.7	56.8	65.6	26.1	3.0
18	180.0	2.3	1188.3	60.9	24.9	105.0	29.8	7.0

shows the moisture content after predrying with infrared radiation (M_{ir}). As it can be seen, the minimum value was obtained in Experiment 14 ($M_{ir} = 59.1$ g/100 g w.b., i.e., a reduction of 24.3%), which was performed at the highest infrared lamp power ($P = 1188.3$ W). The exposure time of the material to the infrared lamps was only 9 min in all runs (Silva et al., 2021).

The average residence time (τ) of the material in the roto-aerated dryer ranged from 3.0 min (Experiment 17) to 6.8 min (Experiment 3), and under the conditions used by Silva et al. (2021) the holdup value was between 65.6 and 194.2 g. The final temperature of acerola residues varied between 33.3°C and 86.3°C. Regarding the final moisture of the material after passing through the roto-aerated dryer (Mroto), it ranged from 65.8 (Experiment 1) to 36.6 g/100 g w.b. (Experiment 14), suggesting that the moisture content of the material was reduced by up to 53.5% in a total time of 13.5 min, including the predrying time. This is a short time compared to other drying techniques already used for dehydrating acerola residues. Nóbrega (2012) showed that the residence time of this material in a tray dryer should be from 120 to 220 min, while Silva et al. (2018) concluded that it takes 159.3–300.7 min for a fixed bed to dry this residue. It is also worth mentioning that prolonged drying times can result in both the degradation of bioactive compounds and the reduction of functional properties (Rehder, 2021; Silva et al., 2020).

The drying of acerola residues in this new hybrid device occurs in a short time due to the synergistic effect of infrared radiation and drying in the roto-aerated dryer. The infrared leads to a rapid heating of the material (Silva et al., 2021), whereas the roto-aerated dryer provides an excellent fluid–particle contact (Silvério et al., 2015) increasing the mass transfer rate and consequently reaching the desired final moisture in a shorter time.

The results reported by Silva et al. (2021) reveal that the operating variables had a significant effect on the drying process. For example, when the drying air temperature was raised from 67.6°C (Experiment 9) to 162.4°C (Experiment 10), the moisture removal increased from 14.7% to 30.6%. Similarly, when the power was increased from 511.7 W (Experiment 13) to 1188.3 W (Experiment 14), the reduction of moisture increased from 16.4% to 42.1%.

Figure 3.13 shows the results of the quantification of bioactive compounds and antioxidant activity after drying acerola residues in each test of the experimental design proposed by Silva et al. (2021) (Experiments 1–18) and those obtained with the fresh residue (line). The total phenolic content (TPC), total flavonoid content (TFC), ascorbic acid content (AA), and antioxidant activity (IC50) are displayed in Figure 3.13a–d, respectively.

As shown in Figure 3.13a, the TPC values were lower at higher infrared lamp powers and/or higher drying air temperatures (Silva et al., 2021). The greatest reductions in TPC were obtained in Experiments 14 and 18 (21.0% and 23.6%, respectively) (Silva et al., 2021). The phenolic losses can be explained by the enzymatic and nonenzymatic oxidative reactions and

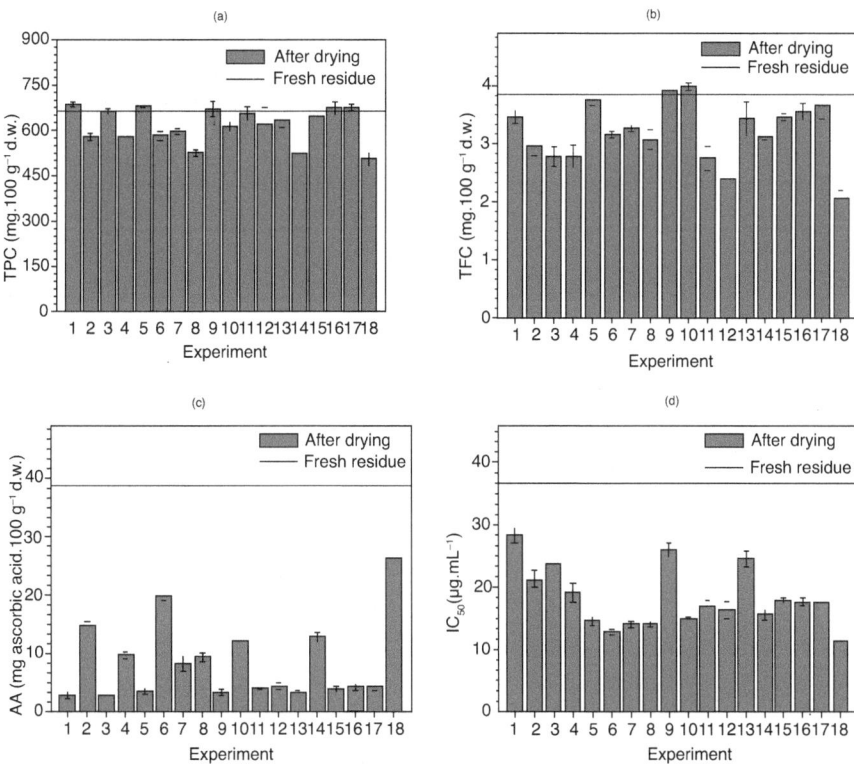

FIGURE 3.13 Total bioactive compounds of acerola residues before (fresh sample) and after drying.

thermal degradation (Nogueira et al., 2019). Figure 3.13a also indicates that in some experiments the TPC values after drying were close to that obtained with the fresh residue. In addition, the dried material presented levels of this bioactive compound that are considered intermediate according to Vasco's classification (Vasco et al., 2008).

Figure 3.13b reveals that the results of TFC had a similar behavior to those of TPC (Silva et al., 2021), that is, in some conditions, the TFC values after drying were also similar to those found in the fresh residue. Değirmencioğlu et al. (2016) emphasized that drying can affect the content of bioactive compounds, causing an increase or a decrease in bioactivity due to physical, chemical, and biochemical changes attributable to the effects of high temperatures, oxygen-rich air, and infrared radiation. Moreover, the binding of phenolic compounds to proteins and changes in chemical structures are other factors related to the loss of total phenolic content (Hernandez-Santos et al., 2015).

Regarding the values of AA after drying, they were lower than those obtained with the fresh residue (Silva et al., 2021). Experiments 6 and 18 not only yielded a high moisture removal but also showed the highest levels of AA. A similar behavior was also observed by Ozgur et al. (2011) during convective drying of green and red peppers. The high retention level of AA in these conditions can be attributed to the inactivation of enzymes responsible for the degradation of ascorbic acid (Ozgur et al., 2011; Dorta et al., 2012; Santos-Sanchéz, 2012). Since AA is considered an indicator of product quality (Podsędek, 2007) and the materials dried under severe conditions showed good retention of ascorbic acid, the dehydration method proposed herein can be considered effective.

The lowest IC50 values, which indicate a higher antioxidant activity (DDPH method), were found in the following tests (Silva et al., 2021): Experiments 6, 7, 8, 10, 14, and 18 (Figure 3.13d). Even though these experimental conditions also led to greater moisture removals (Table 3.4), the material showed a high antioxidant activity after drying. Processing methods may improve the properties of natural antioxidants or induce the formation of new compounds with antioxidant properties. This may happen due to various factors, such as the increase in the antioxidant power of polyphenols in an intermediate state of oxidation, the reduction of sugars, and the formation of Maillard reaction products, which are known to have excellent antioxidant capacity (Madrau et al., 2009).

3.2.2 ROTARY DRYER WITH INERT BED

The drying of pasty materials is a common way of producing powders or granules in chemical and food processing industries (Souza et al., 2019). The drying of pasty materials covered with a thin layer formed on the surface of suspended inert particles is generally used in moving bed dryers, including fluidized and spouted beds (Vieira et al., 2019). The use of conventional fluidized beds with inert particles is limited since good fluidization patterns are achieved preferentially with fine particles, classified as A or B on Geldart's map. Spouted beds have greater use than fluidized bed, as the spouting of coarse particles ensures good fluid dynamic characteristics required for drying (Estiati et al., 2021). The product formed is collected by a cyclone in the form of a thin powder. However, there are many limitations that prevent this technique from becoming competitive for industrial application, since the quantity of air required to keep the spout stable is superior to that used in the drying process itself (Freire et al., 2017).

Conventional flighted rotary dryers are intended for granular materials and may not be used to process pasty materials. Nonetheless, nonconventional configurations of rotary dryers can be employed in this case, including the use of an inert bed inside the drum. The inert bed increases the surface contact between hot air and the pasty material, besides preventing loss of material on the walls and in the dryer structure (Silva et al., 2019b).

In rotary dryers with inert bed (RDIB), the fed pulp coats the inert particles with a thin layer and the drying air removes gradually the pulp moisture by convection (air–material) and conduction (inert particles), transforming the material into a dry film. This film is then transformed into powder due to particle–particle and particle–wall collisions. This powder is elutriated by airflow and separated by a cyclone (Silva et al., 2019b). This alternative configuration of the rotary dryer has some advantages over other devices used for pasty material drying (e.g., fluidized and spouted beds), such as a lower drop pressure and a higher moisture removal efficiency, in addition to an inexistence of problems related to instability or bed agglomeration due to the use of larger inert particles (Silva et al., 2019b). Santos et al. (2022) used the experimental apparatus presented in Figure 3.14 to dry the camu-camu (*Myrciaria dubia*) pulp. Camu-camu (*Myrciaria dubia*) is a native fruit to the Amazon rainforest region with an increased amount of vitamin C (more than any other fruit ever studied). Only a 100g of this fruit may contain more than 5,000 mg of vitamin C, which is around 100 and three times the amount of vitamin C found in oranges or lemons and acerola cherries, respectively.

Experiments and DEM simulations were used by Santos et al. (2022) to analyze the performance of an RDIB toward the drying of the camu-camu pulp. The effects of important process variables on the drying yield, including filling degree, rotational speed, and inert fraction, were quantified. An optimization study was also performed to find the best operating conditions. The quality of the final powder product was also investigated through the analysis of the content of bioactive compounds.

In this nonconventional device, the frequency of the collisions of inert particles is directly related to stages involving layer adhesion and subsequent breaking, playing an important role in the final result of the process. In addition to the work developed by Santos et al. (2022), Silva et al. (2019b) also used DEM simulations to better understand the flow of inert particles inside the drum and consequently the drying behavior of Spirulina platensis microalgae in the RDIB. Figure 3.15 shows some DEM results obtained by Silva et al. (2019b) for the velocity of inert particles at different rotational speeds (RS) and a filling degree (FD) of 15%.

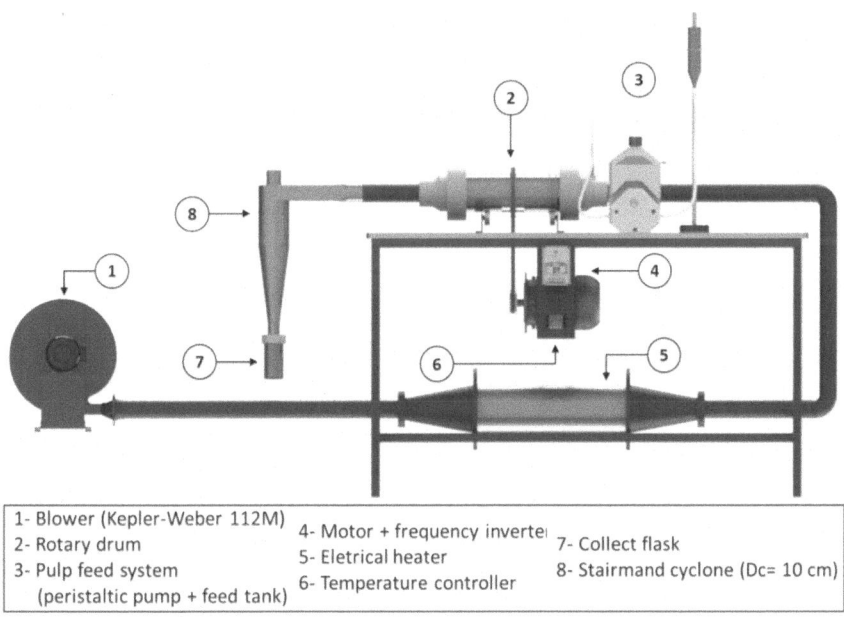

1- Blower (Kepler-Weber 112M)	4- Motor + frequency inverter	7- Collect flask
2- Rotary drum	5- Eletrical heater	8- Stairmand cyclone (Dc= 10 cm)
3- Pulp feed system (peristaltic pump + feed tank)	6- Temperature controller	

FIGURE 3.14   Experimental apparatus used by Santos et al. (2022).

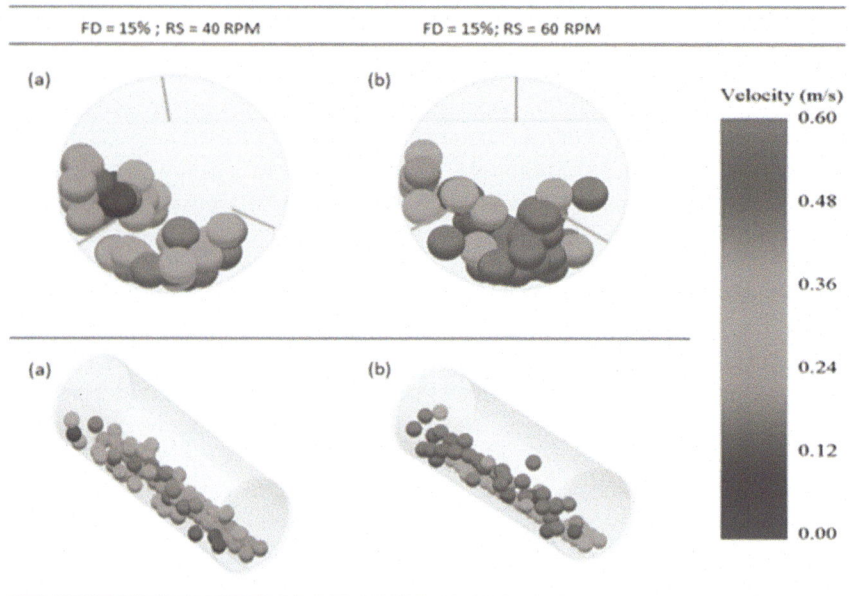

FIGURE 3.15 Velocity of inert particles at the same filling degree (FD) and different rotational speeds (RS).

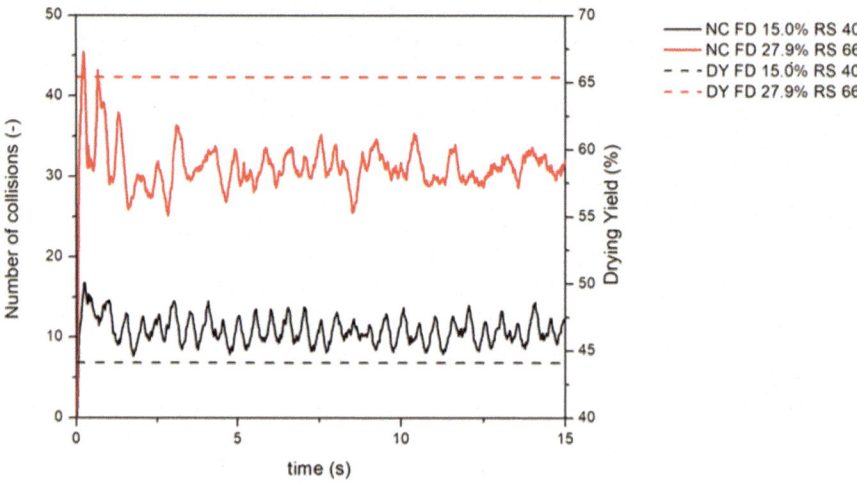

FIGURE 3.16 Number of collisions of inert particles as a function of filling degree (FD) and rotational speed (RS) (2019).

Figure 3.16 shows the DEM results reported by Silva et al. (2019b) for the simultaneous increase in both FD and RS. As it can be observed, when the FD and RS are simultaneously increased, the NC and its magnitude also increase, resulting in a better process performance. According to the figure, the FD and RS values increased by 86% and 65%, respectively, leading to an increment of 186% in the number of collisions and 48.2% in the experimental results of the drying yield (DY). The drying yield is the percentage ratio of the quantity of material collected in the cyclone underflow (dry base) to the quantity of material fed into the system (dry base).

Santos et al. (2022) used porcelain spheres with three different diameters (25.40, 19.05 and 12.70 mm) as inert particles. Six different configurations of inert particles (C1 to C6) were investigated. In the configurations C1, C2, and C3, single diameter particles were used in the inert bed, whereas in the configurations C4, C5, and C6, binary mixtures of particle sizes were used. Figure 3.17 shows the six mixtures of inert particles (Santos et al., 2022).

Figure 3.18 shows the DEM results of the number of collisions (NC) for the six different configurations of inert particles analyzed by Santos et al. (2022) (C1–C6) operating at RS = 60 RPM and FD = 30% for all configurations and FI = 50% for the configurations with binary mixture (C4, C5, and C6). As it can be seen, the greatest number of collisions (NC) was obtained in configuration C3, in which the inert bed was composed of particles with smaller diameter (12.70 mm). The increase in the diameter of inert particles from 12.70 (C3) to 25.40 mm (C1) caused a significant decrease in the number of collisions (14-fold reduction), because the number of particles inside the drum was inversely proportional to their diameter for the same filling degree. A larger number of particles is associated with a greater possibility of collision.

C1	C2	C3	C4	C5	C6
25.40 mm	19.05 mm	12.70 mm	25.40-19.05 mm	25.40-12.70 mm	19.05-12.70 mm

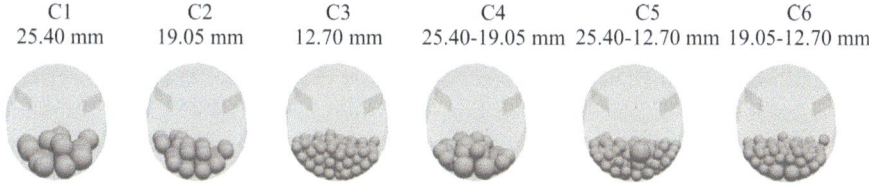

FIGURE 3.17 Drum filling configurations with single (C1, C2, and C3) and binary mixtures (C4, C5, and C6) of inert particles (Santos et al., 2022).

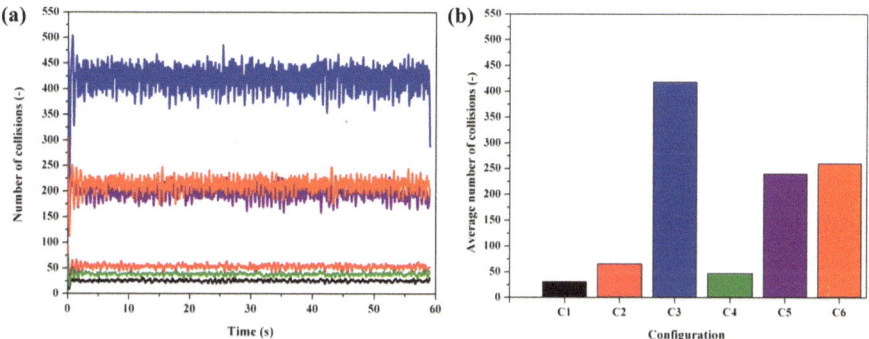

FIGURE 3.18 Number of collisions (NC) over time (a) and mean values (b) in the six configurations of inert bed evaluated (Santos et al., 2022).

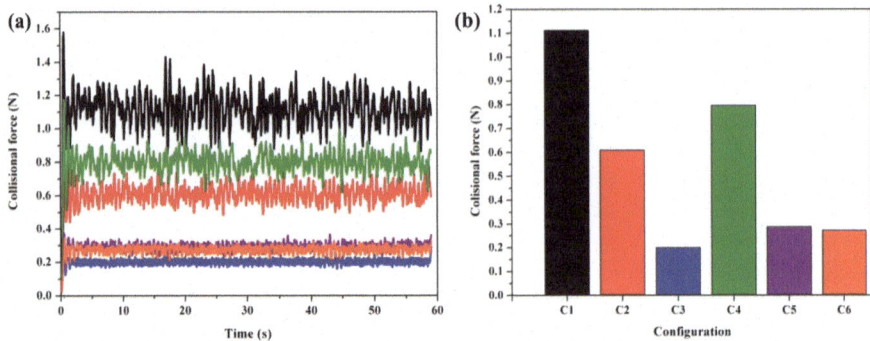

FIGURE 3.19 Force of collision over time (a) and mean values (b) in the six configurations of inert bed evaluated (Santos et al., 2022).

Apart from the number of collisions, the force of collision (FC) among inert particles (intensity of collision) also influences the film removal from these particles. Figure 3.19 shows the DEM results of force of collision (FC) obtained by Santos et al. (2022) for the six inert bed configurations. As demonstrated, the force of collision was directly proportional to the particle diameter, that is, the bigger the inert particle size, the more intense the collision since the mass of particles increases as a function of their diameter.

Santos et al. (2022) studied the effect of operating conditions, i.e., rotational speed (RS), filling degree (FD), and the fraction of inert (FI), on the RDIB performance. FI was defined as the volumetric fraction of the inert particles with the largest diameter in binary mixtures. As previously observed, the number of collisions (NC) and the force of collision (FC) significantly influence the RDIB performance (Silva et al., 2019b). An increase in these responses (NC and FC) promotes a greater removal of the dry film that covers the inert particles, thus increasing the process yield. However, some operating conditions under which the NC is high could also lead to a lower FC due to the opposite effect of FD and FI on both NC and FC. Therefore, it is important to identify the operating conditions that simultaneously maximize the number of collisions and the force of collision. In the work developed by Santos et al. (2022), this condition was identified through a multiresponse optimization study using the desirability function (Derringer and Suich, 1980).

The optimum condition verified by Santos et al. (2022) for the filling degree (FD) was the highest level of FD for all inert bed configurations, i.e., 37.07% for configurations C1, C2, and C3, and 38.41% for configurations C4, C5, and C6 (binary mixtures). These FD conditions led to an increased number of collisions in each configuration evaluated. For granular binary mixtures (C4, C5, and C6), the optimum values of FI led to a higher fraction of large particles, which in turn increased the force of collision. The optimum RS values were different in each inert bed configuration, ranging from 49.21 RPM (C4) to 69.9 RPM (C1).

REFERENCES

Arruda, E.B. Comparison of the performance of the roto-fluidized dryer and conventional rotary dryer, Ph.D. thesis, Federal University of Uberlandia (2008), Uberlandia, Brazil.

Arruda, E.B.; Lobato, F.S.; Assis, A.J.; Barrozo, M.A.S. "Modeling of fertilizer drying in roto-aerated and conventional rotary dryers" *Drying Technology* 27 (2009a): 1192–1198. https://doi.org/10.1080/07373930903263129.

Arruda, E.B.; Façanha, J.M.F.; Pires, L.N.; Assis, A.J.; Barrozo, M.A.S. "Conventional and modified rotary dryer: Comparison of performance in fertilizer drying" *Chemical Engineering and Processing* 48 (2009b): 1414–1418. https://doi.org/10.1016/j.cep.2009.07.007.

Baker, C.G.J. "The design of flights in cascading rotary dryers" *Drying Technology* 6 (1988):631–653. https://doi.org/10.1080/07373938808916402.

Bortolotti, C.T.; Santos, K. G.; Francisquetti, M.C.C.; Duarte, C.R.; Barrozo, M.A.S. "Hydrodynamic study of a mixture of west Indian cherry residue and soybean grains in a spouted bed" *The Canadian Journal of Chemical Engineering* 91 (2013): 1871–1880. https://doi.org/10.1002/cjce.21870.

Britton, P.F.; Sheehan, M.E.; Schneider, P.A. "A physical description of solids transport in flighted rotary dryers" *Powder Technology* 165 (2006): 153–160. https://doi.org/10.1016/j.powtec.2006.04.006.

Chang, C.H.; Lin, H.Y.; Chang, C.Y.; Liu, Y.C. "Comparisons on the antioxidant properties of fresh, freeze-dried and hot-air-dried tomatoes" *Journal of Food Engineering* 77 (2006): 478–485. https://doi.org/10.1016/j.jfoodeng.2005.06.061.

Değirmencioğlu, N.; Gürbüz, O.; Herken, E.N.; Yıldız, A.Y. "The impact of drying techniques on phenolic compound, total phenolic content and antioxidant capacity of oat flour tarhana" *Food Chemistry* 194 (2016). https://doi.org/10.1016/j.foodchem.2015.08.065.

Derringer, G.; Suich, R. "Simultaneous optimization of several response variables" *Journal of Quality Technology* 12 (1980): 214–219. https://doi.org/10.1080/00224065.1980.11980968.

Dorta, E.; Lobo, M.G.; González, M. "Using drying treatments to stabilise mango peel and seed: Effect on antioxidant activity" *LWT-Food Science Technology* 45 (2012): 261–268. https://doi.org/10.1016/j.lwt.2011.08.016.

Duzzioni, A.G.; Lenton, V.M.; Silva, D.I.S.; Barrozo, M.A.S. "Effect of drying kinetics on main bioactive compounds and antioxidant activy of acerola (Malpighia emarginata D. C.) residue" *International Journal of Food Science & Technology* 48 (2013): 1041–1047. https://doi.org/10.1111/ijfs.12060.

Estiati, I.; Atxutegi, A.; Altzibar, H.; Freire, F.B.; Aguado, R.; Olazar, M. "Multiple-output neural network to estimate solid cycle times in conical spouted beds" *Chemical Engineering Technology* 44 (2021): 542–550. https://doi.org/10.1002/ceat.202000491.

Fernandes, N.J.; Ataide, C.H.; Barrozo, M.A.S. "Modeling and experimental study of hydrodynamic and drying characteristics of an industrial rotary dryer" *Brazilian Journal of Chemical Engineering* 26 (2009): 331–341. https://doi.org/10.1590/S0104-66322009000200010.

Freire, F.B.; Atxutegui, A.; Freire, J.T.; Aguado, R.; Olazar, M. "An adaptive lumped parameter cascade model for orange juice solid waste drying in spouted bed" *Drying Technology* 35 (2017): 577–584. https://doi.org/10.1080/07373937.2016.1190937.

Hernandez-Santos, B.; Vivar-Vera, M.A.; Rodriguez-Miranda, J.; Herman-Lara, E.; Torruco-Uco, J.G.; Acevedo-Vendrell, O.; Martinez-Sanchez, C.E. "Dietary fibre and antioxidant compounds in passion fruit (Passiflora edulis f. flavicarpa) peel and depectinised peel waste" *Institute of Food Science Technology* 50 (2015): 268–274. https://doi.org/10.1111/ijfs.12647.

Keey, R.B.; Danckwerts, P.V. *Drying: Principles and Practice*, Pergamon Press Ltd, Oxford, 2013.

Kemp, I.C. "Comparison of particles motion correlations for cascading rotary dryers" in: *Proceedings of the 14th International Drying Symposium*, São Paulo, Brazil, (2004): 790–797.

Lisboa, M.H.; Vitorino, D.S.; Delaiba, W.B.; Finzer, J.R.D.; Barrozo, M.A.S. "A study of particle motion in rotary dryer" *Brazilian Journal of Chemical Engineering*, 24 (2007): 265–374. https://doi.org/10.1590/S0104-66322007000300006.

Madrau, M.A.; Piscopo, A.; Sanguinetti, A.M.; Del Caro, A.; Poiana, M.; Romeo, F.V.; Piga, A. "Effect of drying temperature on polyphenolic content and antioxidant activity of apricots" *European Food Research Technology* 228 (2009): 441. https://doi.org/10.1007/s00217-008-0951-6.

Nascimento, S.M.; Lima, R.M.; Brandão, R.J.; Duarte, C.R.; Barrozo, M.A.S. "Eulerian study of flights discharge in a rotating drum" *Canadian Journal of Chemical Engineering* 97 (2019): 477–484. https://doi.org/10.1002/cjce.23291.

Nóbrega, E.M.M.A. Secagem de resíduo de acerola (Malphigia emarginata DC.): estudo do processo e avaliação do impacto sobre o produto final, Master's thesis, Federal University of Rio Grande do Norte (2012), Brazil.

Nogueira, G.D.R.; Silva, P.B.; Duarte, C.R.; Barrozo, M.A.S. "Analysis of a hybrid packed bed dryer assisted by infrared radiation for processing acerola (Malpighia emarginata D.C.) residue" *Food and Bioproducts Processing* 114 (2019): 235–244. https://doi.org/10.1016/j.fbp.2019.01.007.

Ozgur, M.; Ozcan, T.; Akpinar-Bayizit, A.; Yilmaz-Ersan, L. "Functional compounds and antioxidant properties of dried green and red peppers" *African Journal of Agriculture Research* 6 (2011): 5638–5644. https://doi.org/10.5897/AJAR11.709.

Podsędek, A. "Natural antioxidants and antioxidant capacity of Brassica vegetables: A review" *LWT-Food Science Technology* 40 (2007): 1–11. https://doi.org/10.1016/j.lwt.2005.07.023.

Rehder, A.P.; Silva, P.B.; Xavier, A.M.F.; Barrozo, M.A.S. "Optimization of microwave-assisted extraction of bioactive compounds from a tea blend" *Journal of Food Measurement and Characterization* 15 (2021): 1588–1598. https://doi.org/10.1007/s11694-020-00750-4.

Santos, R.L.; Brandão, R.J.; Nunes, G.; Duarte, C.R.; Barrozo, M.A.S. "Analysis of particles collisions in a newly designed rotating dryer and its impact on the Camu-Camu pulp drying" *Drying Technology* 40 (2022): 2034–2045. https://doi.org/10.1080/07373937.2021.1915795.

Santos-Sánchez, N.F.; Valadez-Blanco, R.; Gómez-Gómez, M.S.; Pérez-Herrera, A.; Salas-Coronado, R. "Effect of rotating tray drying on antioxidant components, color and rehydration ratio of tomato saladette slices" *LWT-Food Science Technology* 46 (2012): 298–304. https://doi.org/10.1016/j.lwt.2011.09.015.

Sheehan, M.E.; Britton, P.F.; Schneider, P.A. "A model for solids transport in flighted rotary dryers based on physical considerations" *Chemical Engineering Science* 60 (2005): 4171–4182. https://doi.org/10.1016/j.ces.2005.02.055.

Silva, D.I.S.; Silva, N.C.; Mendes, L.G.; Barrozo, M.A.S. "Effects of thick-layer drying on the bioactive compounds of acerola residues" *Journal of Food Process Engineering* 41 (2018): e12854. https://doi.org/10.1111/jfpe.12854.

Silva, D.I.S.; Souza, G.F.M.V.; Barrozo, M.A.S. "Heat and mass transfer of fruit residues in a fixed bed dryer: Modeling and product quality" *Drying Technology* 37 (2019a): 1321–1327. https://doi.org/10.1080/07373937.2018.1498509.

Silva, N.C.; Machado, M.V.C.; Brandão, R.J.; Duarte, C.R.; Barrozo, M.A.S. "Dehydration of microalgae Spirulina platensis in a rotary drum with inert bed" *Powder Technology* 351 (2019b): 178–185. https://doi.org/10.1016/j.powtec.2019.04.025.

Silva, N.C.; Santana, R.C.; Duarte, C.R.; Barrozo, M.A.S. "Impact of freeze-drying on bioactive compounds of yellow passion fruit residues" *Journal of Food Process Engineering* 40 (2017): 1–9. https://doi.org/10.1111/jfpe.12514.

Silva, P.B.; Duarte, C.R.; Barrozo, M.A.S. "Dehydration of acerola (Malpighia emarginata D.C.) residue in a new designed rotary dryer: Effect of process variables on main bioactive compounds" *Food and Bioproducts Processing*, 98 (2016): 62–70. https://doi.org/10.1016/j.fbp.2015.12.008.

Silva, P.B.; Mendes, L.G.; Rehder, A.P.B.; Duarte, C.R.; Barrozo, M.A.S. "Optimization of ultrasound-assisted extraction of bioactive compounds from acerola waste" *Journal of Food Science and Technology* 57 (2020): 4627–4636. https://doi.org/10.1007/s13197-020-04500-8.

Silva, P.B.; Nogueira, G.D.R.; Duarte, C.R.; Barrozo, M.A.S. "A new rotary dryer assisted by infrared radiation for drying of acerola residues" *Waste and Biomass Valorization*, 12 (2021): 3395–3406. https://doi.org/10.1007/s12649-020-01222-y.

Silveira, J.C.; Brandao, R.J.; Lima, R.M.; Machado, M.V.C.; Barrozo, M.A.S.; Duarte, C.R. "A fluid dynamic study of the active phase behavior in a rotary drum with flights of two and three segments" *Powder Technology* 368 (2020): 297–307. https://doi.org/10.1016/j.powtec.2020.04.051.

Silveira, J.C.; Lima, R.M.; Brandao, R.J.; Duarte, C.R.; Barrozo, M.A.S. "A study of the design and arrangement of flights in a rotary drum" *Powder Technology* 395 (2022): 195–206. https://doi.org/10.1016/j.powtec.2021.09.043.

Silvério, B.C. Fluid dynamics and drying study on fertilizer drying with different types of rotary dryers, Ph.D. thesis, Federal University of Uberlandia, Uberlandia, Brazil, 2012.

Silvério, B.C.; Arruda, E.B.; Duarte, C.R.; Barrozo, M.A.S. "A novel rotary dryer for drying fertilizer: Comparison of performance with conventional configurations" *Powder Technology* 270 (2015): 135–140. https://doi.org/10.1016/j.powtec.2014.10.030.

Silvério, B.C.; Façanha, J.M.F; Arruda, E.B.; Murata, V.V.; Barrozo, M.A.S. "Fluid dynamics in concurrent rotary dryer and comparison of their performance with a modified dryer" *Chemical Engineering Technology* 34 (2011): 81–86. https://doi.org/10.1002/ceat.201000338.

Sousa, M.S.B.; Vieira, L.M.; Silva, M.J.M.; Lima, A. "Nutritional characterization and antioxidant compounds in pulp waste of tropical fruits" *Agrotechnical Science* 35 (2011): 554–559. https://doi.org/10.1590/S1413-70542011000300017.

Souza, R.C.; Freire, F.B.; Altzibar, H.; Ferreira, M.C.; Freire, J.T. "Drying of pasty and granular materials in mechanically and conventional spouted beds" *Particuology* 42 (2019): 176–183. https://doi.org/10.1016/j.partic.2018.01.006.

Vasco, C.; Ruales, J.; Kamal-Eldin, A. "Total phenolic compounds and antioxidant capacities of major fruits from Ecuador" *Food Chemistry*, 111 (2008): 816–823. https://doi.org/10.1016/j.foodchem.2008.04.054.

Vieira, G.N.A.; Freire, F.B.; Olazar, M.; Freire, J.T. "Real-time monitoring of milk powder moisture content during drying in a spouted bed dryer using a hybrid neural soft sensor" *Drying Technology* 37 (2019): 1184–1190. https://doi.org/10.1080/07373937.2018.1492614.

4 Fluid Dynamics and Modeling of Heat and Mass Transfer in Rotary Drums

4.1 INTRODUCTION

The rotary drum design and prediction of performance are intrinsically related to the ability of estimating all the complex transport phenomena inside this equipment, such as the momentum, including the particle–particle and particle–fluid interactions, and the heat and mass transfer. Besides experimental techniques, numerical simulations can be used to fundamentally investigate the fluid–solid interaction inside rotary drums, in an attempt to connect the material properties and operating conditions with the measured results, consequently removing the empiricism toward predictive design and operation. Two different numerical approaches have been widely applied to predict the particle dynamics in rotary drums: the two-fluid model or Euler–Euler approach, implemented through CFD (computational fluid dynamics) techniques, and the Lagrangian approach, which is implemented by means of the discrete element method (DEM). Furthermore, the complex drying process inside rotary dryers, which simultaneously involves solid transportation and heat and mass transfer, can also be modeled by using simplified models (i.e., distributed parameter and lumped parameter models) based on mass and energy balances for both solid and drying air phases, as well as empirical correlations to estimate the drying kinetics, the equilibrium moisture content of solids, and the heat transfer coefficient.

4.2 CFD SIMULATION AND VALIDATION THROUGH EXPERIMENTS

4.2.1 CFD MODELING

For the fluid–solid multiphase modeling using the Euler–Euler approach, both phases are mathematically treated as interpenetrating continua. Thus, partial differential equations of momentum and mass and energy transfer, along with some constitutive equations, are solved in an Eulerian frame of reference. The conservation of mass for the fluid and granular solid phases is given by Equations (4.1) and (4.2), whereas the corresponding conservation of momentum is obtained by Equations (4.3) and (4.4). Herein, the indexes f and s refer to the fluid and granular solid phases, respectively.

$$\frac{\partial}{\partial t}\left(\rho_f \, \alpha_f\right)+\nabla\cdot\left(\rho_f \, \alpha_f \, \overrightarrow{v_f}\right)=0 \tag{4.1}$$

$$\frac{\partial}{\partial t}\left(\rho_s \, \alpha_s\right)+\nabla\cdot\left(\rho_s \, \alpha_s \, \overrightarrow{v_s}\right)=0 \tag{4.2}$$

$$\frac{\partial}{\partial t}\left(\alpha_f \, \rho_f \, \overrightarrow{v_f}\right)+\nabla\cdot\left(\alpha_f \, \rho_f \, \overrightarrow{v_f} \, \overrightarrow{v_f}\right)$$
$$=-\alpha_f \, \nabla p+\nabla\cdot\overline{\overline{\tau}}_f+\alpha_f\rho_f\vec{g}+\left[\beta\left(\overrightarrow{v_s}-\overrightarrow{v_f}\right)\right] \tag{4.3}$$

$$\frac{\partial}{\partial t}\left(\alpha_s \, \rho_s \, \overrightarrow{v_s}\right)+\nabla\cdot\left(\alpha_s \, \rho_s \, \overrightarrow{v_s} \, \overrightarrow{v_s}\right)$$
$$=-\alpha_s \, \nabla p-\nabla p_s+\nabla\cdot\overline{\overline{\tau}}_s+\alpha_s\rho_s\vec{g}+\left[\beta\left(\overrightarrow{v_f}-\overrightarrow{v_s}\right)\right] \tag{4.4}$$

where $\vec{v}$ is the velocity vector, α is the volume fraction, ρ is the density, p is the thermodynamics pressure, p_s is the total granular solid pressure, $\overline{\overline{\tau}}$ is the viscous stress tensor, $\vec{g}$ is the gravitational acceleration vector, and β is the solid–fluid momentum exchange coefficient, which is related to the drag force.

The fluid and solid viscous stress tensors, $\overline{\overline{\tau}}_f$ and $\overline{\overline{\tau}}_s$, respectively, are modeled according to Equations (4.5) and (4.6). Herein, T and $\overline{I}$ refer to the transpose of a matrix and the identity matrix, respectively.

$$\overline{\overline{\tau}}_f=\alpha_f \, \mu_{\text{eff}}\left[\nabla \, \overrightarrow{v_f}+\left(\nabla \, \overrightarrow{v_f}\right)^T\right]-\frac{2\alpha_f}{3}\mu_{\text{eff}}\left(\nabla\cdot\overrightarrow{v_f}\right)\overline{I} \tag{4.5}$$

$$\overline{\overline{\tau}}_s=\alpha_s \, \mu_s\left[\nabla\overrightarrow{v_s}+\left(\nabla \, \overrightarrow{v_s}\right)^T\right]+\alpha_s\left(\lambda_s-\frac{2}{3}\mu_s\right)\left(\nabla\cdot\overrightarrow{v_s}\right)\overline{I} \tag{4.6}$$

where μ_{eff} is the effective fluid viscosity, which can be related to the fluid dynamic viscosity $\left(\mu_f\right)$ and the turbulent fluid viscosity $\left(\mu_{f,T}\right)$ through $\mu_{\text{eff}}=\mu_f+\mu_{f,T}$, μ_s is the total granular solid viscosity, and λ_s is the bulk granular solid viscosity.

Among many turbulence models in the literature (Menter, 1994), the $k_f-\varepsilon_f$ model, classified as a Reynolds-averaged Navier–Stokes (RANS) equation turbulence model, has been widely used to estimate the turbulent fluid viscosity $\left(\mu_{f,T}\right)$. In this case, the turbulent fluid viscosity is given by $\mu_{f,T}=C_\mu\rho_f k_f^2 / \varepsilon_f$, and k_f and ε_f are calculated from the two following additional transport equations:

$$\frac{\partial}{\partial t}\left(\alpha_f \rho_f k_f\right) + \nabla \cdot \left(\alpha_f \rho_f \overrightarrow{v_f}\, k_f\right)$$

$$= \nabla \cdot \left(\alpha_f \frac{\mu_{fT}}{\sigma_k} \nabla k_f\right) + \alpha_f \left(G - \rho_f \varepsilon_f\right) + \prod_{kw} \tag{4.7}$$

$$\frac{\partial}{\partial t}\left(\alpha_f \rho_f \varepsilon_f\right) + \nabla \cdot \left(\alpha_f \rho_f \overrightarrow{v_f}\, \varepsilon_f\right)$$

$$= \nabla \cdot \left(\alpha_f \frac{\mu_{fT}}{\sigma_\varepsilon} \nabla \varepsilon_f\right) + \alpha_f \frac{\varepsilon_f}{k_f}\left(C_1 G - C_2 \rho_f \varepsilon_f\right) + \prod_{\varepsilon w} \tag{4.8}$$

where k_f is the turbulent kinetic energy, ε_f is the turbulent kinetic energy dissipation rate, $\prod_{kw}$ and $\prod_{\varepsilon w}$ are the dispersed phase turbulence influences on the continuous phase (Elghobashi and Abou-Arab, 1983; Cokljat et al., 2000), and G is the turbulent kinetic energy production rate. The model constants are $C_\mu = 0.09$, $C_1 = 1.44$, $C_2 = 1.92$, $\sigma_k = 1.0$, and $\sigma_\varepsilon = 1.3$ (Launder and Spalding, 1974).

The kinetic theory of granular flow (KTGF), originally proposed by Lun et al. (1984), has been commonly used to model the total granular solid viscosity (μ_s) and the bulk granular solid viscosity (λ_s). Table 4.1 presents the constitutive equations for the granular solid momentum closure.

According to the KTGF, the granular temperature (ψ_s) measures the particle fluctuations in analogy to the thermodynamic temperature (Lun et al., 1984) and is given by the following additional transport equation (Equation 4.22):

$$\frac{3}{2}\left[\frac{\partial}{\partial t}\left(\rho_s \alpha_s \psi_s\right) + \nabla \cdot \left(\rho_s \alpha_s \overrightarrow{v_s} \psi_s\right)\right]$$

$$= \left(-p_s \overline{\overline{I}} + \overline{\overline{\tau}}_s\right) : \nabla \overrightarrow{v_s} + \nabla \cdot \left(k_{\theta s} \nabla \psi_s\right) - \gamma_{\theta s} + \varphi_{fs}. \tag{4.22}$$

Different drag models have been used in the literature to estimate the solid–fluid momentum exchange coefficient (β), such as those developed by Syamlal and O'Brien (1988), Schiller and Naumann (1935), Wen and Yu (1966), Gibilaro et al. (1985), and Huilin et al. (2003), among others.

For the sake of illustration, the Gidaspow drag model (Gidaspow, 1994), which represents a combination between the Ergun model (Ergun, 1952) for $\alpha_f < 0.8$ (i.e., dense systems) and the Wen and Yu model for $\alpha_f \geq 0.8$ (i.e., dilute systems), is described as follows:

TABLE 4.1
Constitutive Equations for the Granular Solid Momentum Closure

- Total granular solid viscosity (μ_s):

$$\mu_s = \mu_{s,kin} + \mu_{s,col} + \mu_{s,fr} \tag{4.9}$$

- Kinetic $(\mu_{s,kin})$ and collisional $(\mu_{s,col})$ granular solid viscosities (Gidaspow, 1994):

$$\mu_{s,col} + \mu_{s,kin} = \frac{4}{5}\alpha_s^2 \rho_s g_{0,ss} d(1+e_s)\sqrt{\psi_s/\pi} + \frac{10 d\rho_s \sqrt{\psi_s \pi}}{96(1+e_s)\alpha_s g_{0,ss}}\left[1 + \frac{4}{5}g_{0,ss}\alpha_s(1+e_s)\right]^2 \tag{4.10}$$

where d, e_s, and ψ_s are the particle diameter, the coefficient of restitution, and the granular temperature, respectively;

- Frictional granular solid viscosity $(\mu_{s,fr})$ (Schaeffer, 1987):

$$\mu_{s,fr} = \frac{p_s \sin(\beta_{fr})}{2\sqrt{I_{2D}}}, \tag{4.11}$$

where β_{fr} and I_{2D} are the frictional internal angle and the second invariant of the deviator of the strain rate tensor, respectively, given as follows:

$$I_{2D} = \frac{1}{6}\left[(D_{Sxx} - D_{Syy})^2 + (D_{Syy} - D_{Szz})^2 + (D_{Szz} - D_{Sxx})^2\right] + D_{Sxy}^2 + D_{Syz}^2 + D_{Szx}^2 \tag{4.12}$$

And

$$D_{Sij} = \frac{1}{2}\left(\frac{\partial v_{si}}{\partial x_j} + \frac{\partial v_{sj}}{\partial x_i}\right) \tag{4.13}$$

- Total granular solid pressure (p_s):

$$p_s = p_{s,kin} + p_{s,col} + p_{s,fr} \tag{4.14}$$

- Kinetic $(p_{s,kin})$ and collisional $(p_{s,col})$ granular solid pressures (Huilin et al., 2003):

$$p_{s,kin} + p_{s,col} = \alpha_s \rho_s \psi_s + 2 \rho_s (1+e_s)\alpha_s^2 g_{0,ss}\psi_s \tag{4.15}$$

- Frictional granular solid pressure $(p_{s,fr})$ (Johnson and Jackson, 1987):

$$p_{s,fr} = 0.05\frac{(\alpha_s - \alpha_{sc})^2}{(\alpha_{s,max} - \alpha_s)^5} \tag{4.16}$$

where α_{sc} and $\alpha_{s,max}$ are the critical solid volume fraction, from which the frictional effects take place, and the maximum packing limit, respectively;

- Radial distribution function $(g_{0,ss})$ (Bagnold, 1954):

$$g_{0,ss} = \left[1 - \left(\frac{\alpha_s}{\alpha_{s,max}}\right)^{\frac{1}{3}}\right]^{-1} \tag{4.17}$$

(Continued)

TABLE 4.1 (*Continued*)
Constitutive Equations for the Granular Solid Momentum Closure

- Bulk granular solid viscosity (λ_s) (Huilin et al., 2003):

$$\lambda_s = \frac{4}{3}\alpha_s^2 \, \rho_s d g_{0,ss} (1+e_s) \sqrt{\frac{\psi_s}{\pi}}$$ (4.18)

- Conductivity of the granular temperature $(k_{\theta s})$ (Syamlal et al., 1993):

$$k_{\theta s} = \frac{150\,\rho_s d\alpha_s \sqrt{\psi_s \pi}}{4(41-33\eta)} \left[1+\frac{12}{5}\eta^2(4\eta-3)\alpha_s g_{0,ss} + \frac{16}{15\pi}(41-33\eta)\eta\alpha_s g_{0,ss}\right]$$ (4.19)

where $\eta = 1/2(1+e_s)$

- Kinetic energy dissipation due to inelastic collisions $(\gamma_{\theta s})$ (Gidaspow, 1994):

$$\gamma_{\theta s} = 3\left(1-e_s^2\right)\alpha_s^2 \rho_s \, g_{0,ss} \, \psi_s \left(\frac{4}{d}\sqrt{\frac{\psi_s}{\pi}} - \nabla \cdot \overrightarrow{v_s}\right)$$ (4.20)

- Kinetic energy dissipation due to fluid friction (φ_{fs}) (Gidaspow, 1994):

$$\varphi_{fs} = -3\beta \, \psi_s$$ (4.21)

$$\beta = \begin{cases} \dfrac{150\alpha_s\left(1-\alpha_f\right)\mu_f}{\alpha_f d^2} + \dfrac{1.75\alpha_s\rho_f\left|\overrightarrow{v_s}-\overrightarrow{v_f}\right|}{d} & \text{if } \alpha_f < 0.8, \\[4mm] \dfrac{3}{4}C_D \dfrac{\alpha_s\alpha_f\rho_f\left|\overrightarrow{v_s}-\overrightarrow{v_f}\right|}{d}\alpha_f^{-2.65} & \text{if } \alpha_f \geq 0.8, \end{cases}$$ (4.23)

where the drag coefficient (C_D) is given by the following equation:

$$C_D = \begin{cases} \dfrac{24}{\alpha_f \, \mathrm{Re}_p}\left[1+0.15\left(\alpha_f \, \mathrm{Re}_p\right)^{0.687}\right] & \text{if } \mathrm{Re}_p < 1000, \\[4mm] 0.44 & \text{if } \mathrm{Re}_p \geq 1000, \end{cases}$$ (4.24)

and the Reynolds dimensionless number (R_{ep}) is calculated as follows:

$$\mathrm{Re}_p = \frac{\rho_f d\left|\overrightarrow{v_s}-\overrightarrow{v_f}\right|}{\mu_f}.$$ (4.25)

Furthermore, the energy equation for the solid and fluid phases is represented by Equation (4.26). Herein, the index i represents either the solid or the fluid phase.

$$\frac{\partial}{\partial t}\left[\alpha_i\,\rho_i\left(\,h_i+k_i\,\right)\right]+\nabla\cdot\left[\alpha_i\,\rho_i\left(\,h_i+k_i\,\right)\vec{v_i}\right]$$

$$=\alpha_i\frac{\partial p}{\partial t}+\nabla\cdot\left(I\alpha_{\text{eff}}\nabla I_i\;\right)+h_T\Delta T, \tag{4.26}$$

where h is the enthalpy, k is the kinetic energy, α_{eff} is the effective thermal diffusivity (i.e., the sum of laminar $\left(\alpha_{L,f}\right)$ and turbulent thermal diffusivities, expressed by $\alpha_{\text{eff}}=\alpha_{L,f}+\mu_{f,T}\,/\left(\rho_f\,\text{Pr}_T\right)$, where $\mu_{f,T}$ is the turbulent fluid viscosity and Pr_T is the turbulent Prandtl dimensionless number), and $h_T\Delta T$ is the convective heat transfer between the involved phases. The convective heat transfer coefficient $\left(h_T\right)$ can be related to the Nusselt dimensionless number (Nu) and given by $h_T=\left(k_f\text{Nu}\right)/d$, where k_f and d are the thermal fluid conductivity and the particle diameter, respectively. Many correlations to estimate the Nusselt number for the convective particle–fluid energy transfer are available in the literature (Ranz and Marshall, 1952; Agarwal, 1988; Larachi et al., 2003; Chen et al., 2021; Huang et al., 2021; Qi and Yu, 2021).

The numerical methods used to solve all the equations presented herein, e.g., discretization schemes, interpolation schemes, and pressure–velocity coupling algorithms, are beyond the scope of this chapter. For a detailed description of the numerical methods commonly applied in CFD simulations, see Versteeg and Malalasekera (2007).

4.2.2 CFD Simulations of Flighted Rotary Drums

A suitable description of the particle–particle (e.g., collisional and frictional effects), particle–fluid (e.g., drag force and turbulence fluctuations), and particle–wall interactions is of fundamental importance when applying the Euler–Euler approach to simulate the particle dynamics in a flighted rotary drum.

Regarding the particle–wall interactions, two different solid phase wall boundary conditions are frequently used: the nonslip wall boundary condition, where all components of the particle velocity on the wall are set to zero, being analogous to the fluid phase, and the Johnson and Jackson (1987) boundary condition, where the particles are allowed to slip on the wall (Zhong et al., 2015). According to the Johnson and Jackson (1987) boundary condition, the particle tangential velocity on the wall can be expressed as follows:

$$u_{s,w}=-\frac{6\alpha_{s,\max}\mu_s}{\sqrt{3}\sqrt{\psi_s}\,\pi\varphi\rho_s\alpha_s g_{0,ss}}\frac{\partial u_{s,w}}{\partial n} \tag{4.27}$$

where $u_{s,w}$ and φ are the tangential particle velocity on the wall and the specularity coefficient, respectively, and n is the normal vector at the wall.

The specularity coefficient $\left(\varphi\right)$ varies from 0 (perfect specular collisions between the particles and the wall) to 1 (perfectly diffuse collisions) and depends on the wall properties (e.g., wall roughness). According to Zhong et al. (2015), although $\varphi=0$

and $\varphi = 1$ are related in the literature to the free-slip and nonslip wall boundary conditions, respectively, there are significant differences in applying the nonslip boundary condition and $\varphi = 1$.

Furthermore, for the simulation of a flighted rotary drum, a dynamic mesh method, in which all computational cells in a specific region of interest in the domain move altogether at the same rotational speed, must be also applied (Jasak, 2009). To analyze the particle–wall boundary condition effect on the particle dynamics in a rotary drum comprised of a single flight, Machado et al. (2017) used different specularity coefficient values in the numerical simulations, including the nonslip boundary condition. The drum had a length of 50.0 cm and an internal diameter of 10.8 cm, whereas the flight had two segments of 1.27 and 0.8 cm, respectively, with an intersegment angle of 90°. Glass beads with a diameter of 1.09 mm were used as particulate material. The end caps of the drum were made of transparent glass to allow particle movement observation. The numerical simulations were validated against experimental data of the solids holdup in the flight at different angular positions, i.e., from 0° to 120°, among other variables.

To measure the experimental solids holdup in the flights, the authors used a high-speed video camera to take pictures of the cross-sectional area occupied by the solids in the flight and ImageJ® software for image analysis. A detailed description of the experimental methodology can be found in Nascimento et al. (2018).

For the sake of illustration, Figure 4.1 shows the experimental and simulated solids holdup in the flight at different angular positions, i.e., from 0° to 120°. In this case, the rotational speed of the rotary drum was 36.1 rpm (Machado et al., 2017). It can be noted (Figure 4.1) that the nonslip boundary condition did not necessarily correspond to $\varphi = 1$, as also observed by Zhong et al. (2015). According to Machado et al. (2017), in general, $\varphi = 0.5$ better represented the solids holdup in the flight when compared to the experimental data in the case of a rotary drum with a single flight and under the operating conditions employed. Other variables, such as the solids unloading profile of the flights and the height of the bed formed by excess material in the drum were also successfully simulated by the authors.

Depending on the operating conditions and the rotary drum geometry configuration, the turbulence fluctuations of the gas properties should be taken into consideration in the modeling of particle dynamics in a flighted rotary drum, e.g., the $k_f - \varepsilon_f$ model previously shown in Equations (4.7) and (4.8).

Nascimento et al. (2021) performed numerical simulations through the Eulerian approach to evaluate the effects of the presence of gas turbulence fluctuations on the prediction of the passive and active phase behaviors in a flighted rotary drum under different operating conditions and using different number of flights (i.e., 12 and 15 L-shaped flights). The prediction of the solids holdup in the flights using the $k_f - \varepsilon_f$ turbulence model (Equations 4.7 and 4.8) was also analyzed.

Figure 4.2 shows the experimental and simulated (i.e., with and without a turbulence model) particle distributions at the end caps (12 L-shaped flights) of the drum under different operating conditions, where the simulated passive (Figure 4.2a) and active (Figure 4.2b) phases can be clearly observed. For the simulated active phase visualization (Figure 4.2b), a suitable solid volume fraction contour filter was applied (Nascimento et al., 2021).

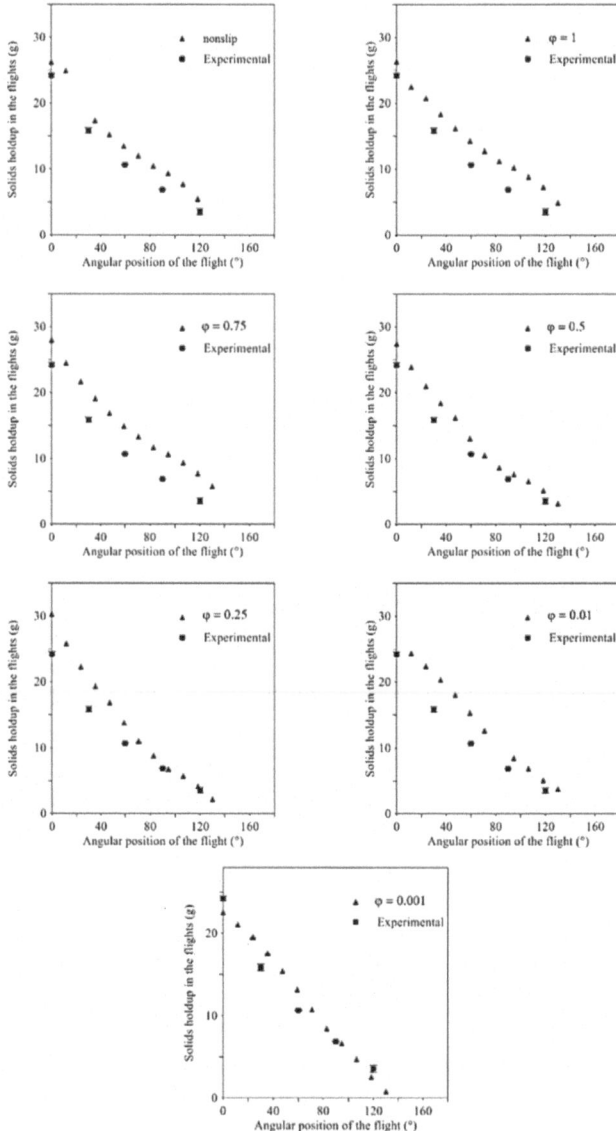

FIGURE 4.1 Effect of specularity coefficient (φ) on the simulated solids holdup in the flights as a function of angular position for a drum rotating at 36.1 rpm (Machado et al., 2017).

According to the authors, regardless of the operating conditions, the simulations using the turbulence model better qualitatively represented the active and passive phase behaviors when compared to the experiments. A quantitative comparison between the experimental and simulated (with and without a turbulence model) solids holdup in the flights (12 L-shaped flights) at different angular positions and under different operating conditions is shown in Figure 4.3.

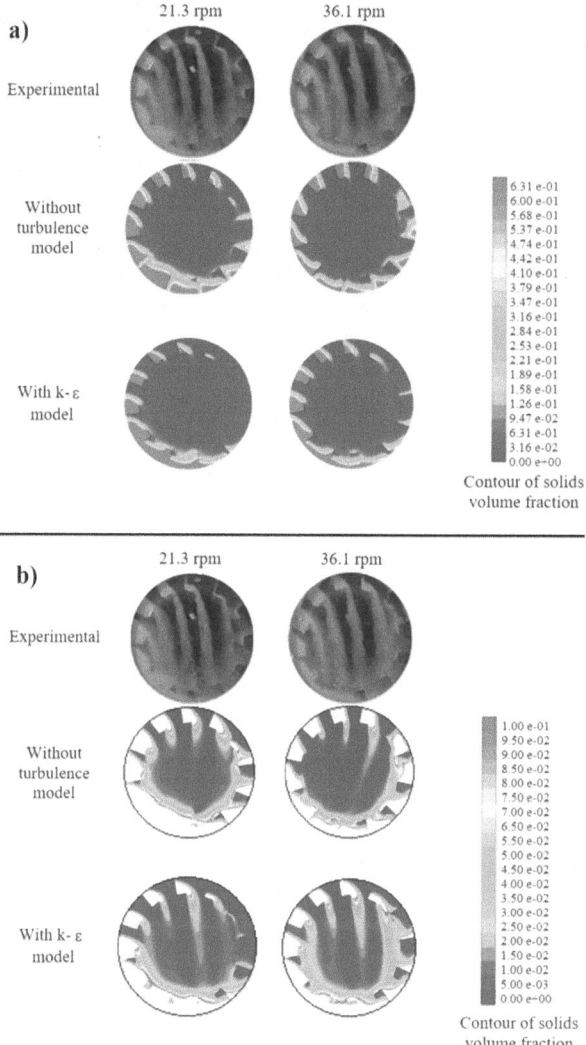

FIGURE 4.2 Experimental and numerical particle distribution at the end caps (12 L-shaped flights) of the drum under different operating conditions (filling degree of 20%): (a) simulations highlighting the passive phase and (b) simulations highlighting the active phase using a solid volume fraction contour filter (Nascimento et al., 2021).

Average percentage deviations of 12.3% and 27.3% were reported between the experiments and the simulations with and without a turbulence model, respectively. The complete discharge of particles from the flights observed in the experiments and predicted through numerical simulations using the $k_f - \varepsilon_f$ turbulence model occurred at about 120° and 130° for the drum rotating at 21.3 and 36.1 rpm, respectively (Nascimento et al., 2021). In addition, the authors demonstrated that the numerical

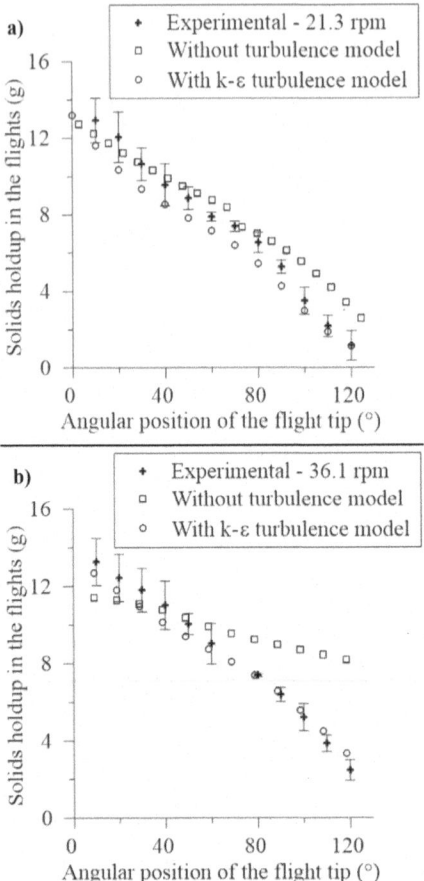

FIGURE 4.3 Experimental and numerical solids holdup in the flights (12 L-shaped flights) at different angular positions and under different operating conditions (filling degree of 20%): (a) drum rotating at 21.3 rpm and (b) drum rotating at 36.1 rpm (Nascimento et al., 2021).

simulations using the $k_f - \varepsilon_f$ turbulence model were also capable of predicting the particle dynamics for a drum operating with 15 L-shaped flights, as shown in Figure 4.4.

With respect to particle–particle interactions, the frictional effects are shown to be fundamental for modeling dense systems (i.e., regions under high solid concentrations) and must be taken into consideration when calculating the granular solid viscosity and granular solid pressure as well as the kinetic and collisional effects, as previously shown through Equations (4.9), (4.11), (4.14), and (4.16) (Table 4.1). The frictional effect takes place whenever $\alpha_s \geq \alpha_{sc}$, where α_s and α_{sc} are the solid volume fraction and the critical solid volume fraction for the frictional effects, respectively, whose value is commonly assumed to be about 0.5.

In the case of a drum operating in a batch mode, where no fluid enters or leaves the drum and the relative velocity between the phases is low, the drag force can usually

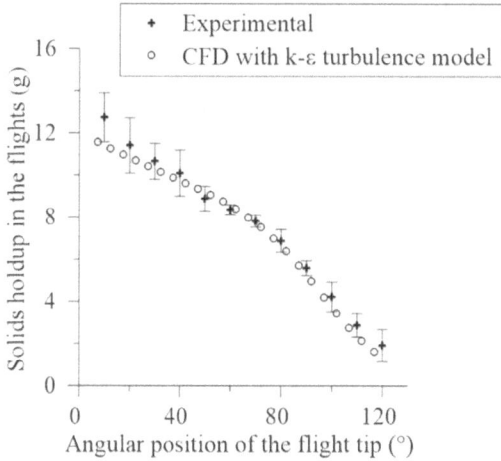

FIGURE 4.4 Experimental and numerical solids holdup in the flights (15 L-shaped flights) at different angular positions: filling degree of 20% and rotational speed of 21.3 rpm (Nascimento et al., 2021).

be neglected. Additionally, since in the Euler–Euler approach the particle shape is not directly considered in the model, except in the drag effects (Ullah et al., 2019), some authors have proposed a relationship between particle shape (e.g., particle sphericity) and α_{sc} to model the nonspherical particle dynamics inside rotary drums without flights using this approach (Benedito et al., 2018; Benedito et al., 2022). One of the advantages of DEM simulations over Euler–Euler CFD simulations is that in the former the particle shape can be directly taken into consideration, as will be discussed in the next sections. Up to now, studies applying the Euler–Euler approach to flighted rotary drums, mainly when operated in a continuous mode, are still scarce in the literature.

4.3 DEM SIMULATION AND VALIDATION THROUGH EXPERIMENTS

4.3.1 DEM Modeling

Originally proposed by Cundall and Strack (1979), the DEM applies a certain force–displacement law to the particle–particle and particle–boundary contacts, thus updating the unbalanced contact forces and Newton's second law to track each individual particle in a Lagrangian framework, considering the body forces acting on it.

The linear and angular momentum equations to calculate the motion of an individual particle i are given by Equations (4.28) and (4.29), respectively:

$$m_i \frac{d\vec{v}_i}{dt} = \sum_{j=1}^{N_c} \left(\overline{F_n^{ij}} + \overline{F_t^{ij}} \right) + m_i \vec{g} \qquad (4.28)$$

$$I_i \frac{d\vec{\omega}_i}{dt} = \sum_{j=1}^{N_c} \left(\vec{\tau_{t,ij}} + \vec{\tau_{r,ij}} \right) \tag{4.29}$$

where m_i is the mass of a particle i, $\vec{v}_i$ is the linear velocity vector of a particle i, I_i is the moment of inertia of a particle i, $\vec{\omega}_i$ is the angular velocity vector of a particle i, $\vec{g}$ is the gravitational acceleration vector, and N_c is the total number of particles in contact with particle i. Furthermore, $\vec{F_n^{ij}}$ is the normal contact force, $\vec{F_t^{ij}}$ is the tangential contact force, $\vec{\tau_{t,ij}}$ is the tangential torque, and $\vec{\tau_{rij}}$ is the rolling torque between particles i and j or particles i and wall boundaries.

To model both particle–particle and particle–boundary collisions, two different contact approaches are commonly used: the hard-sphere contact model (Alder and Wainwright, 1957), where the collisions are assumed to be instantaneous and the binary collisions dominant (e.g., rapid flows of dilute systems) and the soft-sphere contact model (Cundall and Strack, 1979), where the particles are allowed to overlap during the collision time (i.e., rigid spheres), which is not instantaneous but can take multiple time steps, and the overlapping magnitude is used to calculate the elastic and frictional forces. The hard-sphere contact model is simpler than the soft-sphere contact model since in the former the physical details of contacts are not taken into consideration (Luding, 2008). Furthermore, the soft-sphere contact model is the most used to calculate particle contact forces in dense systems (such as those found in rotary drums).

The soft-sphere contact model is based on the spring-dashpot model (Cundall and Strack, 1979) illustrated in Figure 4.5, where k, λ, μ_f, and δ are the stiffness coefficient (which represents the spring effect), the damping coefficient (which represents the dashpot effect), the coefficient of static friction, and the overlap distance between the involved particles, respectively. The indexes n and t refer to the normal and tangential directions, respectively.

For spherical particles, the overlap distance can be defined as $\delta = R_i + R_j - d$, where R_i, R_j, and d are the radii of particles i and j and the distance between the center of mass of both particles, respectively. Therefore, δ corresponds to the maximum penetration between the particles.

According to the spring-dashpot model, the normal $\left(\vec{F_n^{ij}} \right)$ and tangential $\left(\vec{F_t^{ij}} \right)$ forces can be decomposed into a conservative spring force term $\left(\vec{F_n^{ij}}^S \text{ and } \vec{F_t^{ij}}^S \right)$ and a dissipative damping force term $\left(\vec{F_n^{ij}}^D \text{ and } \vec{F_t^{ij}}^D \right)$ as follows:

$$\vec{F_n^{ij}} = \vec{F_n^{ij}}^S + \vec{F_n^{ij}}^D \tag{4.30}$$

$$\vec{F_t^{ij}} = \vec{F_t^{ij}}^S + \vec{F_t^{ij}}^D \tag{4.31}$$

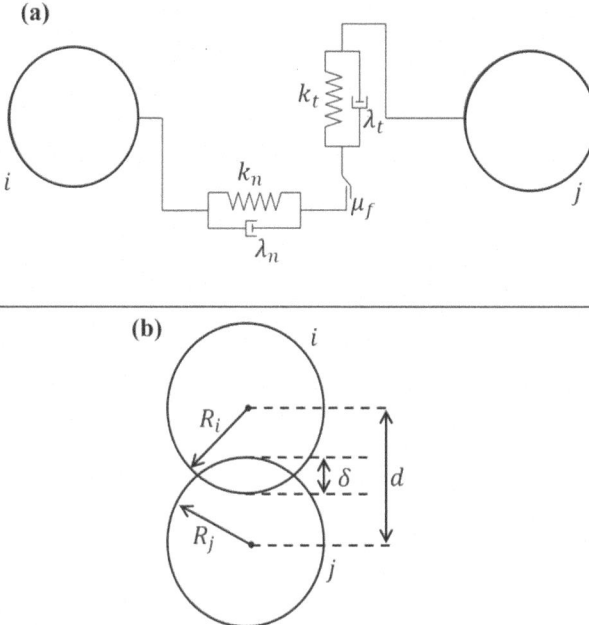

FIGURE 4.5 Schematic illustration showing (a) the spring-dashpot model and (b) the particle–particle overlap (soft-sphere model).

Depending on the expressions used for the stiffness and damping terms, the soft-sphere contact model can be classified into linear and nonlinear contact model. The liner contact model or linear spring-dashpot model is based on Hooke's law (Caserta et al., 2016), whereas different nonlinear models have been proposed in the literature (Hertz, 1882; Mindlin and Deresiewicz, 1953; Walton and Braun, 1986; Langston et al., 1994; Di Maio and Di Renzo, 2005).

Among all contact models, the nonlinear Hertz–Mindlin model, where the normal and tangential contact forces are based on the Hertzian contact theory and the Mindlin model, respectively, has been frequently applied in DEM simulations (Paulick et al., 2015; Zhang et al., 2020). In the Hertz–Mindlin model, the normal and tangential conservative spring forces are given by Equations (4.32) and (4.33), respectively:

$$\overrightarrow{F_n^{ij}}^S = -k_n \delta_n \vec{n}_{ij} = -\frac{4}{3} E^* \sqrt{R^* \delta_n}\, \delta_n \vec{n}_{ij} \tag{4.32}$$

$$\overrightarrow{F_t^{ij}}^S = -k_t \delta_t \vec{t}_{ij} = -8 G^* \sqrt{R^* \delta_n}\, \delta_t \vec{t}_{ij} \tag{4.33}$$

where $\vec{n}_{ij}$ and $\vec{t}_{ij}$ are the normal and tangential unit vectors, respectively. The equivalent Young's modulus (E^*) (Equation 4.34), equivalent shear modulus (G^*) (Equation 4.35), and equivalent radius (R^*) (Equation 4.36) are defined as follows:

$$\frac{1}{E^*} = \frac{1-v_i^2}{E_i} + \frac{1-v_j^2}{E_j} \tag{4.34}$$

$$\frac{1}{G^*} = \frac{2(2-v_i)(1+v_i)}{E_i} + \frac{2(2-v_j)(1+v_j)}{E_j} \tag{4.35}$$

$$\frac{1}{R^*} = \frac{1}{R_i} + \frac{1}{R_j} \tag{4.36}$$

where v is Poisson's ratio of particle i or j.

On the other hand, the normal and tangential dissipative damping forces are calculated by Equations (4.37) and (4.38), respectively:

$$\overrightarrow{F_n^{ij}}^{D} = \lambda_n v_n^{\overline{rel}} = 2\sqrt{\frac{5}{6}}\beta\sqrt{\left(2E^*\sqrt{R^*\delta_n}\right)m^*}\, v_n^{\overline{rel}} \tag{4.37}$$

$$\overrightarrow{F_t^{ij}}^{D} = \lambda_t v_t^{\overline{rel}} = 2\sqrt{\frac{5}{6}}\beta\sqrt{\left(8G^*\sqrt{R^*\delta_n}\right)m^*}\, v_t^{\overline{rel}} \tag{4.38}$$

where the equivalent mass (m^*) (Equation 4.39) and the damping ratio (β) (Equation 4.40) are defined as follows:

$$\frac{1}{m^*} = \frac{1}{m_i} + \frac{1}{m_j} \tag{4.39}$$

$$\beta = \frac{\ln(e_s)}{\sqrt{\ln^2(e_s)+\pi^2}} \tag{4.40}$$

where e_s, $v_n^{\overline{rel}}$, and $v_t^{\overline{rel}}$ are the coefficient of particle restitution and the normal and tangential relative velocities between particles i and j, respectively.

The tangential force $\left(\overline{F_t^{ij}}\right)$ in Equation (4.31) is limited by Coulomb's friction law:

$$\overline{F_t^{ij}} = -\mu_f\left|\overline{F_n^{ij}}\right|\vec{t}_{ij}, \quad if \quad \left|\overline{F_t^{ij}}\right| > \mu_f\left|\overline{F_n^{ij}}\right| \tag{4.41}$$

Finally, the tangential torque $\left(\overrightarrow{\tau_{t,ij}}\right)$ and the rolling torque $\left(\overrightarrow{\tau_{rij}}\right)$ between particles i and j in Equation (4.29) can be represented through Equations (4.42) and (4.43), respectively:

$$\overrightarrow{\tau_{t,ij}} = \vec{R}_{ij} \times \overrightarrow{F_t^{ij}} \tag{4.42}$$

$$\overrightarrow{\tau_{rij}} = -\mu_r \left| k_n \delta_n \right| R^* \frac{\hat{\omega}_{t,ij}}{\left| \hat{\omega}_{t,ij} \right|} \tag{4.43}$$

where $\vec{R}_{ij}$ is the vector from the center of particle i to j, μ_r is the coefficient of rolling friction, and $\hat{\omega}_{t,ij}$ is the relative angular velocity vector between particles i and j.

4.3.2 DEM SIMULATIONS OF FLIGHTED ROTARY DRUMS

DEM simulations involve a set of physical parameters used to represent the elasticity of particles and wall boundaries (i.e., the spring and dashpot effects) as well as the energy loss due to frictional effects during particle–particle and particle–wall interactions. The spring and dashpot effects are related to the stiffness (k) and damping (λ) coefficients, respectively, and can be obtained through Poisson's ratio (v), Young's modulus (E), the shear modulus (G), and the coefficient of restitution (e_s) of particles and wall boundaries, as previously shown in Equations (4.32)–(4.40), whereas the frictional effects are associated with the coefficient of static friction (μ_f) and the coefficient of rolling friction (μ_r) of particle–particle and particle–wall contacts (Equations 4.41 and 4.43).

Since the DEM codes usually apply an explicit scheme to integrate the balance force equations over time, the time increments (i.e., the time step Δt used in DEM simulations) must be kept smaller than a critical value (Δt_{crit}) to avoid numerical instabilities (Otsubo et al., 2017). Two different criteria for the critical time step calculation are commonly defined for the simulation of particle dynamics in a flighted rotary drum: Rayleigh's criterion, based on the propagation time of a Rayleigh wave from one pole to another in a particle, and Hertz's criterion, based on the particle–particle collision time.

The collision time based on Hertz's criterion ($\Delta t_{crit, Hertz}$) can be estimated as follows (Zhang et al., 2020):

$$\Delta t_{crit,Hertz} = 2.87 \left(\frac{m^{*2}}{R^* E^{*2} V_{c,max}} \right)^{0.2} \tag{4.44}$$

where $V_{c,max}$ is the maximum collision velocity.

Zhang et al. (2020) assumed a $V_{c,\max}$ equal to $2\,\mathrm{m\,s^{-1}}$ to simulate the particle dynamics in a flighted rotary drum operating at a maximum rotational speed of 30 rpm and found a $\Delta t_{\mathrm{crit,\,Hertz}}$ of about 2.99×10^{-4}·s for plastic balls with a diameter of 6 mm and a bulk density of 596 kg·m⁻³. To guarantee numerical stability, the authors considered a time step of 10^{-5} s, which is less than 20% of the critical time step ($\Delta t < 0.2 \times \Delta t_{\mathrm{crit,\,Hertz}}$) frequently recommended in the literature (Danby et al., 2013).

On the other hand, taking into consideration the physical properties of the smallest particles in the system, e.g., particle i, Rayleigh's criterion $\left(\Delta t_{\mathrm{crit,Rayleigh}}\right)$ can be expressed as follows (Silveira et al., 2022):

$$\Delta t_{\mathrm{crit,Rayleigh}} = \frac{\pi R_i \sqrt{\rho_i\,/\,G_i}}{0.1613 \upsilon_i + 0.8766} \tag{4.45}$$

For glass beads with a diameter of 1.09 mm and a density of 2455 kg·m⁻³, Silveira et al. (2020, 2022) calculated a critical Rayleigh time step of 5.7×10^{-5} s. However, for the sake of numerical stability, they used a time step of 10^{-6} s, which is 1.7% of Rayleigh's criterion, to simulate the particle motion inside a flighted rotary drum.

To avoid extremely small time steps in DEM simulations when calculating either $\Delta t_{\mathrm{crit,Hertz}}$ or $\Delta t_{\mathrm{crit,Rayleigh}}$, the particles are frequently considered to be less stiff than they are in reality. Thus, a reduction in either Young's modulus (E) or the shear modulus (G) by two to three orders of magnitude was shown not to significantly affect the overall simulation behavior (Zhang et al., 2020; Silveira et al., 2020; Zhang et al., 2021; Silveira et al., 2022).

Two different strategies have been widely used to determine the DEM interaction parameters (i.e., coefficient of restitution $\left(e_s\right)$, coefficient of static friction (μ_f), and coefficient of rolling friction $\left(\mu_r\right)$) in the simulation of flighted rotary drums: a direct experimental measurement at the particle level and a calibration procedure focused on the bulk level (Coetzee, 2017; Coetzee, 2020).

Regarding the direct experimental measurements, many suitable experimental apparatuses have been developed, as illustrated in Figure 4.6.

The coefficient of restitution $\left(e_s\right)$, which represents the energy conservation after a particle–particle or a particle–wall collision, can be measured as the ratio of the relative velocities between two objects after and before their collision. The e_s value varies from 0 (perfect inelastic collision) to 1 (perfect elastic collision) (Zhang et al., 2021). If the frictional forces are neglected and the particle is dropped from rest against a horizontal surface, Equation (4.46) can be applied, where h_2 and h_1 are the particle bounce height and the particle drop height, respectively (Batista et al., 2021; Lima et al., 2021). To capture the collision events, a high-speed video camera is frequently employed (Figure 4.6a).

$$e_s = \sqrt{\frac{h_2}{h_1}} \tag{4.46}$$

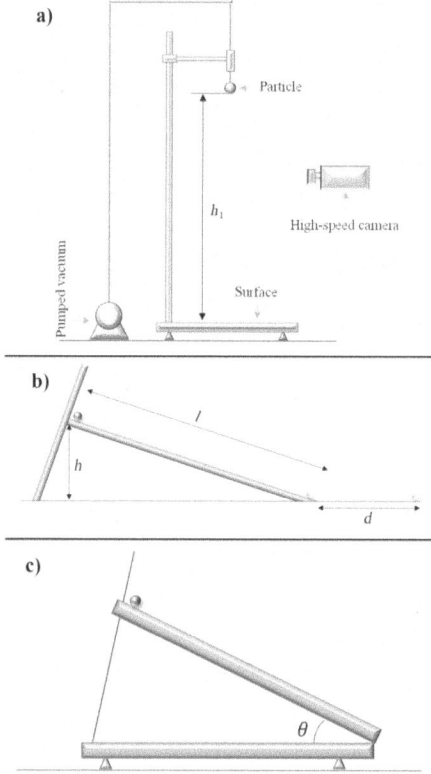

FIGURE 4.6 Schematic illustration of some apparatuses used to measure the DEM interaction parameters (Batista et al., 2021): (a) coefficient of restitution, (b) coefficient of rolling friction, and (c) coefficient of static friction.

An additional mirror can be used during particle drop to guarantee a particle trajectory in the vertical axis (Lima et al., 2021). Furthermore, Jiang et al. (2020) developed a methodology to measure the coefficient of restitution of nonspherical particles (i.e., irregular particles) by the particle tracking velocimetry (PTV) method.

To measure the coefficient of static friction (μ_f), an apparatus comprised of an inclined plane is usually employed (Figure 4.6c), where the angle of inclination is increased until the particles, initially positioned over the plane, start to slide. To avoid particle rolling instead of sliding, the particles are usually glued together in a group of three or four. Additionally, to measure the coefficient of static friction for the particle–particle interaction, the corresponding particles are frequently glued onto the plane as a layer (Batista et al., 2021; Lima et al., 2021). In this case, Equation (4.47) is used:

$$\mu_f = \tan\theta \tag{4.47}$$

where θ represents the sliding angle.

Finally, the coefficient of rolling friction (μ_r), which is defined as the ratio of the frictional force to the normal force that prevents the rolling movement of a particle, can be measured according to the ASTM standards (ASTM G194–08, 2018) by means of an angled launch ramp (Figure 4.6b). As it can be seen in Figure 4.5b, μ_r can be calculated through Equation (4.48):

$$\mu_r = \frac{h}{d} \qquad (4.48)$$

where d is the total distance traveled by the particle after leaving the ramp and h is the vertical ramp height.

On the other hand, the calibration procedure to determine the DEM parameters are based on the experimental measurement of a specific bulk property, e.g., the static and dynamic angles of repose of the bed material, and a subsequent comparison with the numerical results, as illustrated in Figure 4.7.

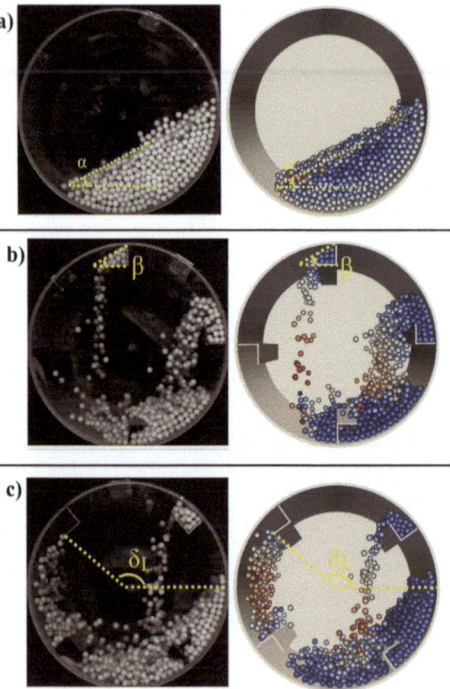

FIGURE 4.7 Illustration of the calibration procedure to determine the coefficients of friction through a comparison between experimental and DEM simulated results of bulk particle property: (a) particle–particle and particle–wall interactions and (b) particle–flight interaction (Zhang et al., 2020).

In this case, the DEM parameters are iteratively changed until the numerical simulation matches the experiments to a certain degree of accuracy (Coetzee, 2017). Most of the works about the calibration procedure use a design of experiment technique to analyze the effects of changing the DEM parameters on the desired bulk response (Santos et al. 2016; Nascimento et al., 2021). According to Coetzee (2017), the calibration procedure has the disadvantage of presenting, in some cases, more than one suitable set of parameter values that well represent a specific particle bulk property. Additionally, a suitable set of parameter values determined for a certain experiment might not accurately represent the bulk behavior in another experiment or application.

Both procedures, direct experimental measurements and calibration, are often combined to determine suitable DEM parameter values for the simulations of particle dynamics in flighted rotary drums (Zhang et al., 2020; Silveira et al., 2020; Zhang et al., 2021; Nascimento et al., 2021; Silveira et al., 2022).

As previously mentioned, contrary to the Euler–Euler approach (CFD), the particle shape can be directly taken into account in DEM simulations. Among many methods, the multisphere and superquadric models are the most popular. As shown in Figure 4.8, the multisphere model approximates the particle shape by connecting a certain number of rigid spheres into a clump particle, whereas the superquadric model represents the particle shape through a smooth surface (Coetzee, 2017; Jiang et al., 2020).

Depending on the real particle shape, the number of particles required to build up a clump (multisphere model) is extremely higher, which increases the computational costs (Santo et al., 2016; Coetzee, 2017). In contrast, the superquadric model uses a continuous function to smoothly represent irregular particles, as shown in Equation (4.49) (Jiang et al., 2020; Wang et al., 2021a):

$$\left(\left| \frac{x}{a} \right|^{n_2} + \left| \frac{y}{b} \right|^{n_2} \right)^{n_1/n_2} + \left| \frac{z}{c} \right|^{n_1} - 1 = 0 \tag{4.49}$$

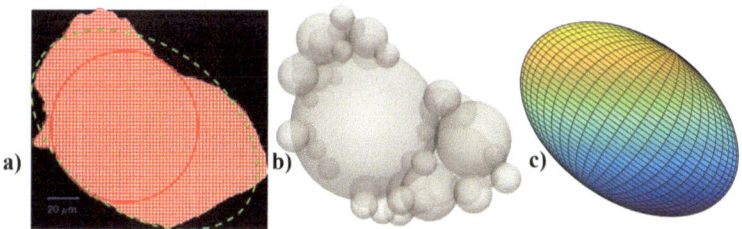

FIGURE 4.8   Illustration of the DEM particle shape representation (irregular particles): (a) scanning electron microscopy (SEM) of a maltodextrin particle (experiment); (b) DEM multisphere model (clump) of the corresponding maltodextrin particle; and (c) DEM superquadric model (ellipsoid) based on the SEM analysis of the corresponding maltodextrin particle (Jiang et al., 2020).

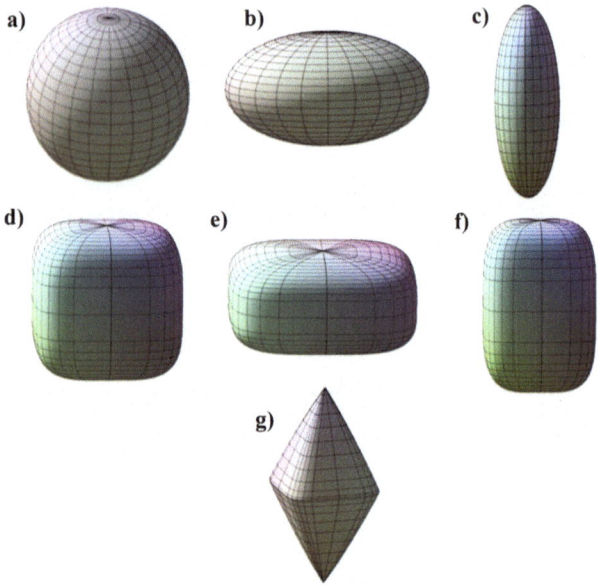

FIGURE 4.9 Illustration of different particle shapes built up using the superquadric function (Equation 4.49): (a) $a : b : c = 1 : 1 : 1$ and $n_1 : n_2 = 2 : 2$; (b) $a : b : c = 2 : 2 : 1$ and $n_1 : n_2 = 2 : 2$; (c) $a : b : c = 2 : 1 : 3$ and $n_1 : n_2 = 2 : 2$; (d) $a : b : c = 2 : 2 : 2$ and $n_1 : n_2 = 4 : 4$; (e) $a : b : c = 2 : 2 : 1$ and $n_1 : n_2 = 4 : 4$; (f) $a : b : c = 2 : 2 : 3$ and $n_1 : n_2 = 4 : 4$; and (g) $a : b : c = 1 : 1 : 2$ and $n_1 : n_2 = 1 : 5$.

where a, b, and c are the half-lengths of the particle along the x, y, and z directions, respectively, and n_1 and n_2 are known as blockiness parameters, which are related to the superquadric element shape. Figure 4.9 depicts different particle shapes as a function of the parameters shown in Equation (4.49).

A detailed description of the algorithm and the contact models for the multisphere and superquadric approaches can be found in Lu et al. (2015), Wang et al. (2021b), and Podlozhnyuk et al. (2017), among other authors.

After calibrating all DEM parameters and validating the subsequent model against experimental data, the model is then suitable for design and optimization processes, substantially decreasing the costs regarding experimental operation and necessary feedstock. Figures 4.10–4.13 demonstrate the application of a validated DEM to investigate the rotary drum geometric effects on the particle behavior.

A very good illustration of the application of DEM simulations to perform process optimization is given by Silveira et al. (2022), who used a multiresponse optimization through the experimental design technique (Box et al., 1978) and the desirability function technique (Derringer and Suich, 1980) in a flighted rotary drum with twelve three-segmented flights. The authors aimed to maximize the solids holdup in the flight (SH), the last unloading flight (LUF), the mass of solids in the active region (SM), and the percentage of the active region occupied by the particles (ARO), in

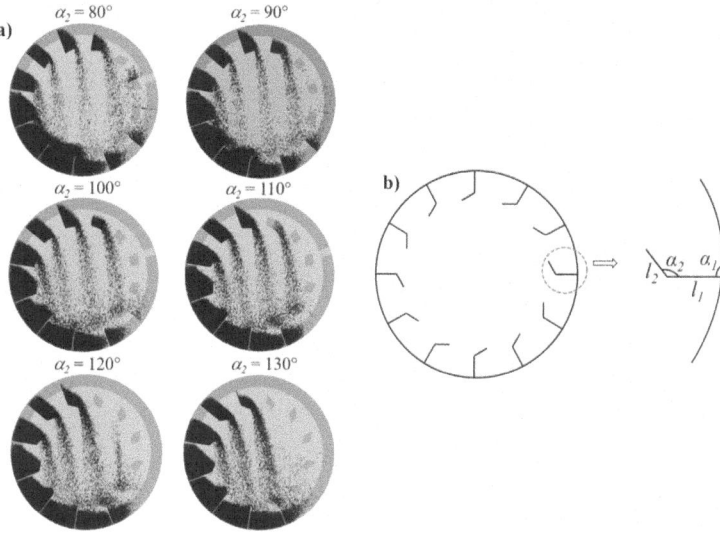

FIGURE 4.10 Simulated particle distributions in a rotary drum (filling degree of 15%; drum rotational speed of 21.3 rpm) using DEM (a) for two-segmented flights and different angulations and (b) with α_1, l_1 and l_2 being kept constant at 90°, 12.7 mm, and 8.0 mm, respectively (Silveira et al., 2020).

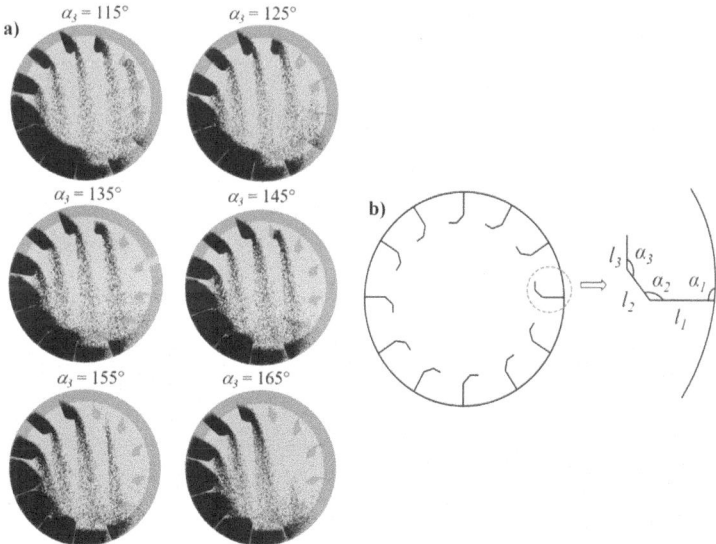

FIGURE 4.11 Simulated particle distributions in a rotary drum (filling degree of 15%; drum rotational speed of 21.3 rpm) using DEM (a) for three-segmented flights and different angulations and (b) with α_1, α_2, l_1, l_2, and l_3 being kept constant at 90°, 135°, 10.0 mm, 4.0 mm, and 4.0 mm, respectively (Silveira et al., 2020).

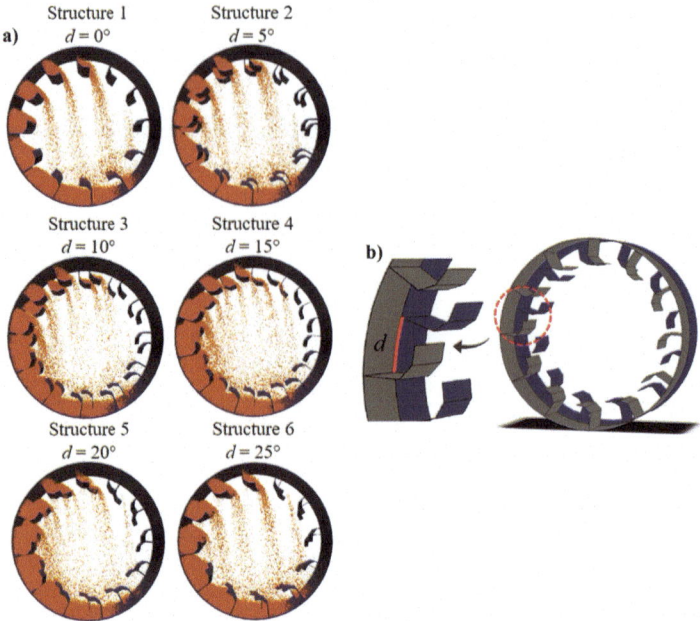

FIGURE 4.12 Simulated particle distributions in a rotary drum (filling degree of 15%; drum rotational speed of 21.3 rpm) using DEM (a) for aligned and interspersed flights and (b) for three-segmented flights with α_1, α_2, and α_3 being kept constant at 95°, 130°, and 145°, respectively (Silveira et al., 2022).

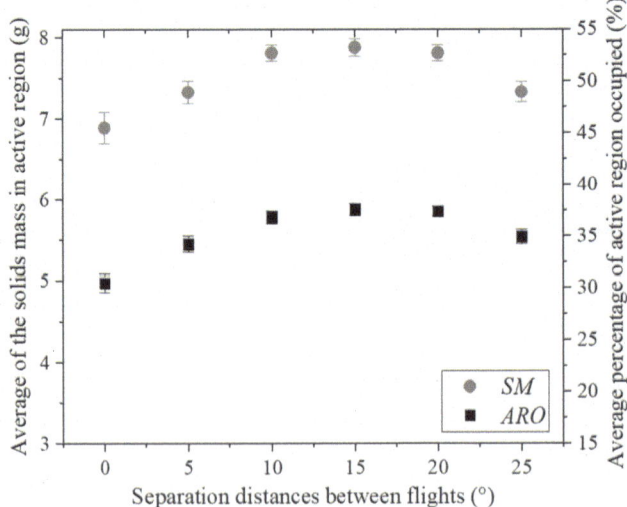

FIGURE 4.13 Simulated mass of solids in the active region (SM) and percentage of the active region occupied by the particles (ARO) for aligned and interspersed flights using different separation distances between flights (d) according to Figure 4.12 (Silveira et al., 2022).

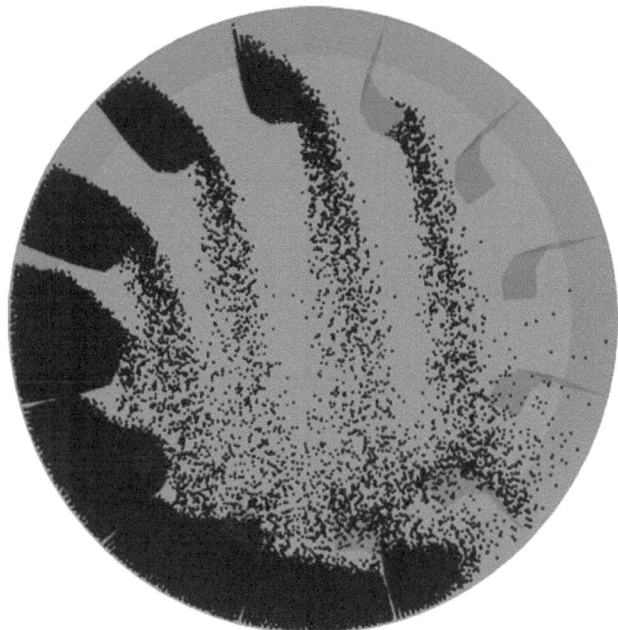

FIGURE 4.14 Simulated particle distribution in a rotary drum (filling degree of 15%; drum rotational speed of 21.3 rpm) using DEM under an optimized condition of angles between segments: $\alpha_1 = 95°$, $\alpha_2 = 130°$, and $\alpha_3 = 145°$ (Silveira et al., 2022).

addition to minimizing the dispersion heterogeneity of solids in the transverse section of the drum (DHT), by varying the angles between segments. The dimensions of the flight segments were kept constant as follows: $l_1 = 10\,mm$ and $l_2 = l_3 = 5\,mm$.

According to the authors, the optimum combination of the angles between segments was $\alpha_1 = 95°$, $\alpha_2 = 130°$, and $\alpha_3 = 145°$. The simulated particle distribution under the optimum condition is shown in Figure 4.14.

4.4 HEAT AND MASS TRANSFER MODELING IN A ROTARY DRYER

Three main transport phenomena simultaneously occur during the drying process inside rotary dryers: solid transportation through passive and active phases along the rotary dryer, heat transfer from the drying air to the wet solid material, and mass transfer (i.e., water vapor) from the wet solid material to the drying air.

A general equation that represents the variation of temperature, air humidity, and solid moisture content over time (t) and along the axial direction (z) in a rotary dryer is as follows:

$$\frac{\partial \xi_i (z,t)}{\partial t} \pm v_i(t) \frac{\partial \xi_i (z,t)}{\partial z} = f_i\left(\xi_i, z, t\right) \tag{4.50}$$

where ξ can be the temperature, air humidity, or solid moisture content (i.e., dependent variable), v is the velocity, which is negative for countercurrent flow and positive for concurrent flow, and f is a function related to either mass or heat transfer. The index i refers to the solid or gas phase.

Depending on the operating conditions and the rotary dryer configuration, Equation (4.50) can be simplified into either distributed parameter or lumped parameter model, which will be discussed next.

4.4.1 DISTRIBUTED PARAMETER MODEL

The distributed parameter model can be applied to long rotary dryers operating under a steady state condition where the drying air humidity, solid moisture content, and temperature gradients are not negligible. Based on Figure 4.15, corresponding to the mass and energy balances for both the solid and drying air phases and considering the assumptions that the solid shape does not change during drying, the particle velocity through the drum is constant, the dryer is operating in countercurrent flow, the drying kinetics takes place in the falling rate period, and the initial conditions (i.e., air and solid temperature, air humidity, and solid moisture content), the solid feeding rate and other parameters are known, Equations (4.51)–(4.54) can be obtained (Arruda et al., 2009; Fernandes et al., 2009; Silva et al., 2012; Souza et al., 2021).

$$\frac{dY_f(z)}{\partial z} = -\frac{R_w H^*}{G_f} \tag{4.51}$$

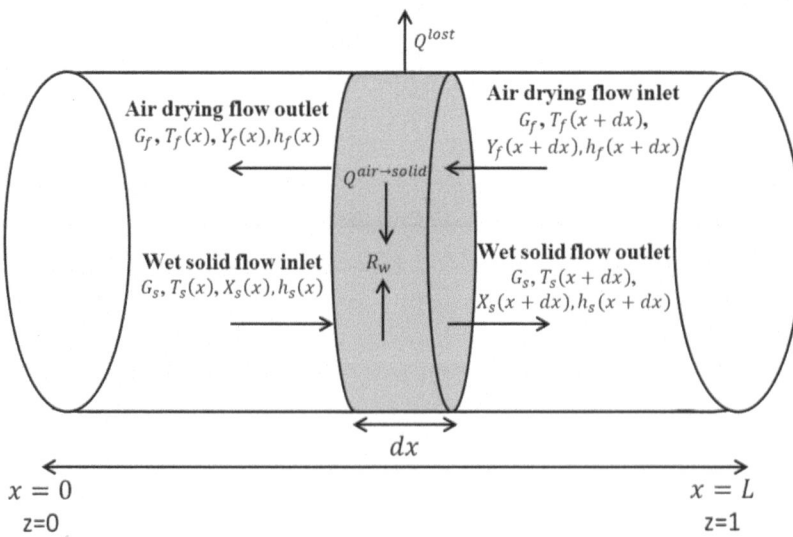

FIGURE 4.15 Infinitesimal volume element of a rotary dryer for the distributed parameter model: dryer operating in countercurrent flow.

$$\frac{dX_s(z)}{\partial z} = -\frac{R_w H^*}{G_s} \tag{4.52}$$

$$\frac{dT_f(z)}{\partial z} = \frac{U_v aV\left[T_f(z) - T_s(z)\right] + R_w H^*\left[\lambda + C_{pv}T_f(z)\right] + U_p \pi DL\left[T_f(z) - T_a\right]}{G_f\left[C_{pf} + Y_f(z)C_{pv}\right]} \tag{4.53}$$

$$\frac{dT_s(z)}{\partial z} = \frac{U_v aV\left[T_f(z) - T_s(z)\right] + R_w H^* C_{pl}T_s(z) - R_w H^*\left\{\lambda + C_{pv}\left[T_f(z) - T_s(z)\right]\right\}}{G_s\left[C_{ps} + X_s(z)C_{pl}\right]}$$

$$\tag{4.54}$$

where Y_f, X_s, R_w, G_s, G_f, T_s, T_f, T_a, $U_v a$, U_p, V, λ, $C_{p(l,\,s,\,\text{and }v)}$, D, L, and z are, respectively, the air absolute humidity (kg$_{\text{water}}$/kg$_{\text{dry air}}$), the solid moisture content (kg$_{\text{water}}$/kg$_{\text{dry solid}}$), the drying rate (min^{-1}), the mass flow rate of solids (kg·min^{-1}), the mass flow rate of the drying air (kg·min^{-1}), the solid temperature (°C), the drying air temperature (°C), the ambient temperature (°C), the volumetric heat transfer coefficient (W·m^{-3}·K^{-1}), the heat transfer coefficient for heat loss from the control volume (W·m^{-2}·K^{-1}), the volume of dryer control volume (m^3), the latent heat of vaporization of water (kJ·kg^{-1}), the specific heat of liquid water, solid, and water vapor (kJ·kg^{-1}·K^{-1}), respectively, the dryer diameter (m), the dryer length (m), and the dimensionless axial position along the dryer ($z = x/L$).

H^* is the total solid load in the dryer (kg) and can be determined by the average particle residence time (τ_R), i.e., $H^* = \tau_R \times G_s$. Since the drying process in a flighted rotary dryer mainly occurs during particle fall from the flights due to a more effective contact between the drying air and the particles, the drying rate is commonly calculated based on a fraction of the particle residence time, which is known as the effective contact time (Arruda et al., 2009b).

Arruda et al. (2009b) applied the distributed parameter model for fertilizer drying in a rotary dryer equipped with three-segmented angular flights using an appropriate set of constitutive equations (see Section 4.4.3). This choice was justified by the fact that this type of flight promotes a more homogeneous cascade through the transversal section of the rotary dryer, enhancing the material spreading and consequently the contact between the falling particles and the drying air. The flights had the following dimensions: first segment with a diameter of 0.02 m and second and third segments with a diameter of 0.007 m; flight length of 1.5 m; and an angle of 135° between segments. The particles used were simple super-phosphate granules (SSPG) with a Sauter mean diameter of 2.45 mm, a particle density of 1100 kg m^{-3}, and a heat capacity of 0.245 kcal·kg^{-1}·°C^{-1}.

The drying process in conventional cascading rotary dryers occurs mostly while the particles are falling from the flights, that is, when they are in contact with the drying air, which corresponds to only a fraction of the residence time. This fraction refers to the effective contact time between the solid and the drying air (f_{tef}) and can be evaluated by the relation between the average falling time of the particle and the total time of a cycle, which in turn corresponds to the time spent from the material collection by the flight to its return to the particle bed at the bottom of the drum. Arruda et al. (2009) evaluated this fraction of residence time using Equation (4.55), where the total number of cycles (N_{Ci}) is given by Equation (4.56) and the effective gas–particle contact time (t_{ef}) is given by Equation (4.57). The total solid load in the dryer (H^*) can be calculated according to Equation (4.58):

$$f_{tef} = \frac{\overline{t_q}}{t_{Ci}} \times \frac{N_{Ci}}{N_{Ci}} = \frac{N_{Ci}\overline{t_q}}{\overline{\tau}} \tag{4.55}$$

$$N_{Ci} = \frac{L}{l} = \frac{L}{\overline{Y_q}\,\mathrm{sen}(\alpha)} \tag{4.56}$$

$$t_{ef} = f_{tef} \times \overline{\tau} \tag{4.57}$$

$$H^* = \overline{\tau} \times G_S \tag{4.58}$$

Figures 4.16–4.18 show typical results obtained by Arruda et al. (2009) from the comparisons between experimental profiles and those computed by the model under specific operating conditions. A good agreement between the simulated and experimentally obtained profiles can be observed in these results, with a slight overprediction of the solid temperature. A good agreement between these profiles was also observed in other conditions studied by these authors.

As already mentioned, for conventional rotary dryers, an effective gas–particle contact time (t_{ef}) substitutes the time t in the drying rate calculation. In contrast, for

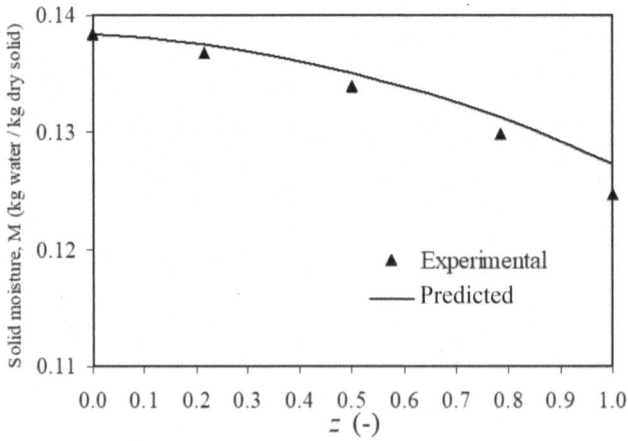

FIGURE 4.16 Experimental and simulated results of solid moisture profile for a rotary dryer operating in the following conditions: inlet air velocity of $1.5\,\mathrm{m\,s^{-1}}$, inlet air temperature of $75\,°C$, and mass flow rate of solids of $1.2\,\mathrm{kg\,min^{-1}}$.

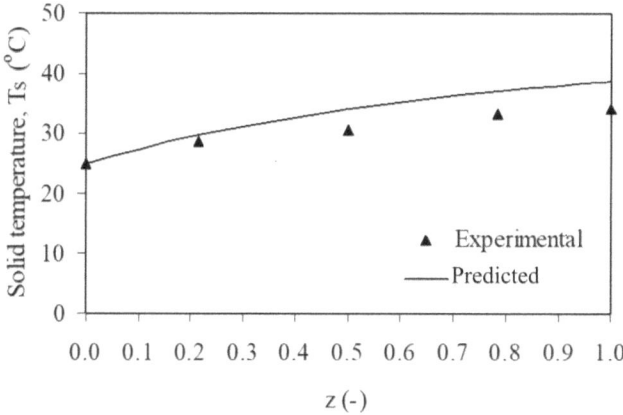

FIGURE 4.17 Experimental and simulated results of solid temperature for a rotary dryer operating in the following conditions: inlet air velocity of $1.5\,m\,s^{-1}$, inlet air temperature of $75\,°C$, and mass flow rate of solids of $1.2\,kg\,min^{-1}$.

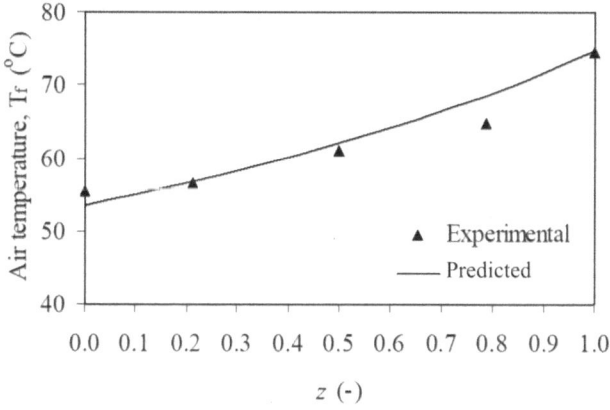

FIGURE 4.18 Experimental and simulated results of air temperature for a rotary dryer operating in the following conditions: inlet air velocity of $1.5\,m\,s^{-1}$, inlet air temperature of $75°C$, and mass flow rate of solids of $1.2\,kg\,min^{-1}$.

roto-aerated dryers, described in Chapter 3, this effective contact time is equal to the residence time itself, that is, the fraction of effective time (f_{tef}) is equal to one.

Typical results of the experimental data and those computed by the distributed parameter model for fertilizer drying in a roto-aerated dryer are shown in Figures 4.19–4.21 for solid moisture distribution and solid and air temperatures along the length of the roto-aerated dryer, respectively. A good agreement between the simulated and experimental results was observed in these experiments and in others performed by Arruda et al. (2009).

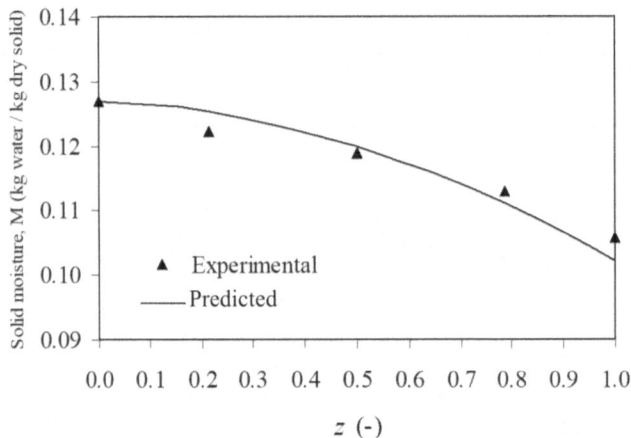

FIGURE 4.19 Experimental and simulated results of solid moisture profile for a roto-aerated dryer operating in the following conditions of test 3: $v_{AR} = 1.5\,\text{m s}^{-1}$, $T_f = 75°C$, and $G_{SU} = 0.8\,\text{kg min}^{-1}$.

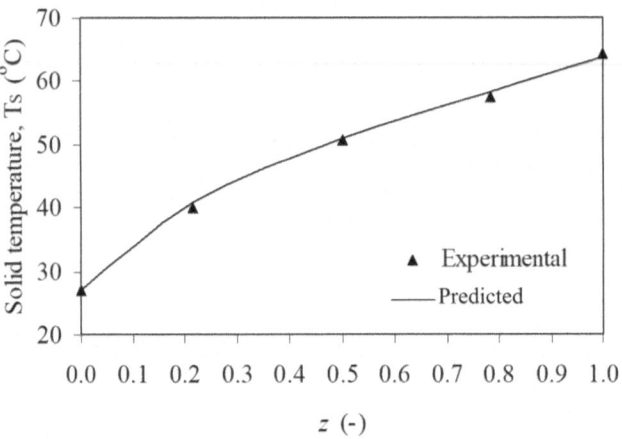

FIGURE 4.20 Experimental and simulated results of solid temperature profile for a roto-aerated dryer operating in the following conditions of test 3: $v_{AR} = 1.5\,\text{m s}^{-1}$, $T_f = 75°C$, and $G_{SU} = 0.8\,\text{kg min}^{-1}$.

4.4.2 LUMPED PARAMETER MODEL

For a dynamic drying analysis (i.e., transient condition), a simple lumped parameter model can be used. In this case, the dryer is usually split into a specific number of subdomains, with each of them being considered a perfectly stirred tank with negligible

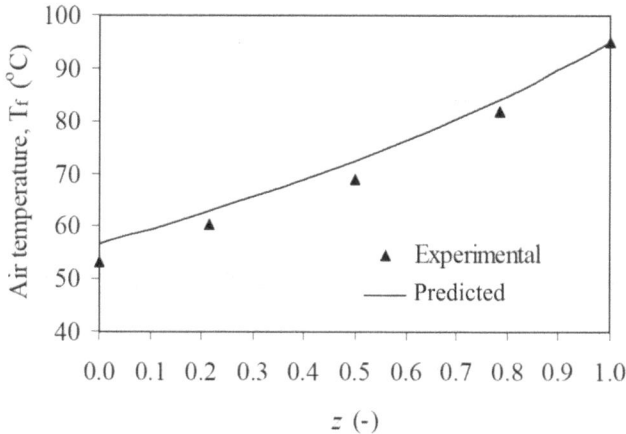

FIGURE 4.21 Experimental and simulated results of air temperature profile for a roto-aerated dryer operating in the following conditions of test 3: $v_{AR} = 1.5\,\text{m s}^{-1}$, $T_f = 75°C$, and $G_{SU} = 0.8\,\text{kg min}^{-1}$.

gradients of air humidity, solid moisture content, and temperature (Douglas et al., 1993; Iguaz et al., 2003).

Figure 4.22a shows a schematic representation of the subdomains for a dryer operating in concurrent flow, whereas Figure 4.22b highlights one subdomain (i.e., control volume) for the mass and energy balances. Each subdomain can also represent one solid recirculation (i.e., equivalent to 1 solids residence time τ) in a short dryer (i.e., with negligible gradients) that needs to recycle the solids to achieve a desired final solid moisture content (Perazzini et al., 2021).

By applying the mass and energy balances to the gas and solid phases in each subdomain (Figure 4.22b) and considering that the gas behaves as an ideal gas with no significant gradients of temperature, solid moisture content, and gas humidity inside each subdomain, the drying kinetics takes place in the falling rate period and the solid shape does not change during the drying process, Equations (4.59)–(4.62) can be obtained as follows:

$$\frac{dX_s(t)}{\partial t} = -\frac{v_s}{L}\left[X_s(t) - X_s^{\text{in}}\right] - R_w \tag{4.59}$$

$$\frac{dY_f(t)}{\partial t} = -\frac{v_f}{L}\left[Y_f(t) - Y_f^{\text{in}}\right] + \frac{H^*}{M_f}R_w \tag{4.60}$$

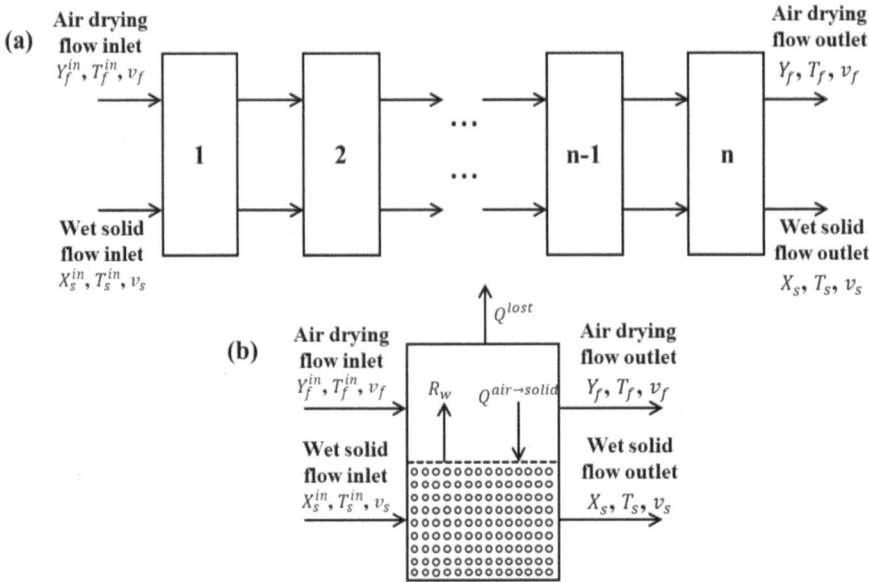

FIGURE 4.22 (a) Rotary dryer control volume and (b) representation of sections for the lumped parameter model: dryer operating in concurrent flow.

$$\frac{dT_s(t)}{\partial t} = -\frac{v_s}{L}\left[T_s(t) - T_s^{in}\right]$$

$$+ \frac{1}{C_{ps} + X_s(t)C_{pl}}\left\{\frac{U_v aV\left[T_f(t) - T_s(t)\right]}{H*} - R_w C_{pv}\left[T_f(t) - T_s(t)\right] - \lambda R_w\right\} \quad (4.61)$$

$$\frac{dT_f(t)}{\partial t} = -\frac{v_f}{L}\left[T_f(t) - T_f^{in}\right]$$

$$+ \frac{1}{C_{pf} + Y_f(t)C_{pv}}\left\{\begin{array}{c}\dfrac{-U_v aV\left[T_f(t) - T_s(t)\right]}{M_f} + R_w C_{pv}T_f(t) + \\ \\ -U_p\pi DL\left[T_f(t) - T_a\right]\end{array}\right\} \quad (4.62)$$

where v_s, v_f, and M_f are the solid velocity (m·s⁻¹), the drying air velocity (m·s⁻¹), and the drying air mass (kg), respectively.

The drying air mass (M_f) can be modeled as follows (ideal gas):

$$M_f = \left(V - \frac{H*}{\rho_s}\right)\frac{M_{wf}M_{ww}}{M_{wf}Y_f + M_{ww}}\frac{P}{R(T_f + 273.15)} \quad (4.63)$$

where

$$M_{wf} = M_{wa}\left(1-Y_f\right)+ M_{ww}Y_f \tag{4.64}$$

where M_{wf}, M_{ww}, M_{wa}, and R are the drying air molecular weight, the water molecular weight, the ambient air molecular weight, and the ideal gas constant, respectively. In the case of the lumped parameter model, the variable H^* in Equations (4.60) and (4.61) is related to the residence time in every control volume (i.e., $H^* = \tau_r \times G_s$) shown in Figure 4.22, which is equal to the residence time in the dryer (τ_R) divided by the total number of control volumes the dryer is divided into (i.e., $\tau_r = \tau_R/n$).

For the sake of illustration of the lumped parameter model, Figures 4.23–4.25 show the experimental and simulated solid temperature, gas temperature, solid moisture content, and gas absolute humidity over the time, respectively (Perazzini et al., 2021). In this case, citrus residues (*Citrus limon L. Burm. f.*) with an apparent specific mass of 1.086 g·cm⁻³, a real specific mass of 3.326 g·cm⁻³, a specific heat of 0.996 J·g⁻¹·K⁻¹, and an initial moisture content of 82% were dried in a rotary dryer with a length of 2.7 m and a diameter of 0.45 m. Six 2.6 m long flights with three segments with lengths of 80 mm for the first segment and 15 mm for the second and third

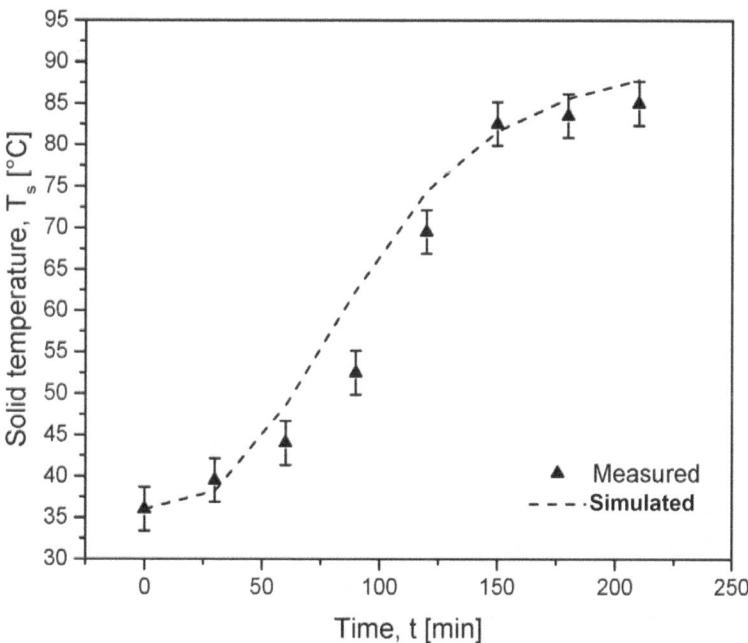

FIGURE 4.23 Experimental and simulated results of solid temperature over time for a rotary dryer operating in the following conditions: mass flow rate of solids of 0.4 kg min⁻¹, inlet air velocity of 2.0 m s⁻¹, inlet air temperature of 155°C, initial solid moisture content of 3.2868 [kg H₂O kg⁻¹ dry material], and inlet gas absolute humidity of 0.0692 [kg H₂O kg⁻¹ dry air] (Perazzini et al., 2021).

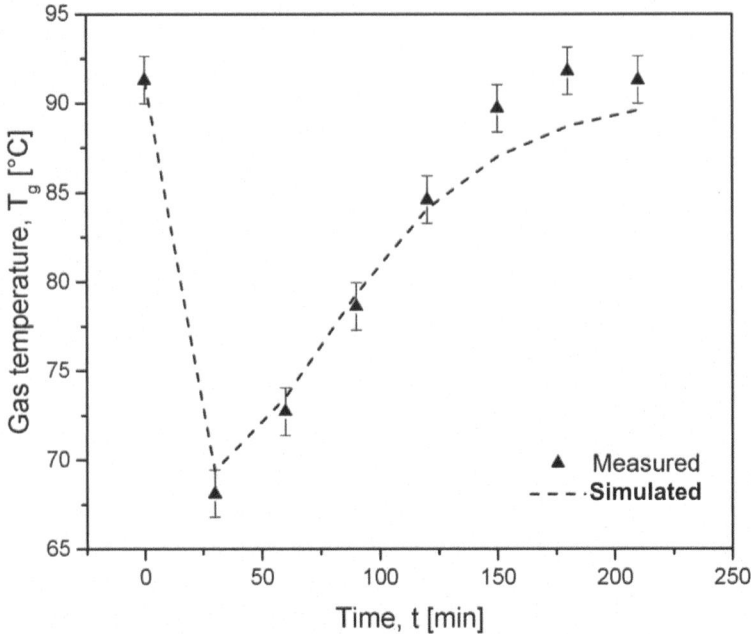

FIGURE 4.24 Experimental and simulated results of gas temperature over time for a rotary dryer operating in the following conditions: mass flow rate of solids of 0.4 kg min⁻¹, inlet air velocity of 2.0 m s⁻¹, inlet air temperature of 155 °C, initial solid moisture content of 3.2868 [kg H_2O kg⁻¹ dry material], and inlet gas absolute humidity of 0.0692 [kg H_2O kg⁻¹ dry air] (Perazzini et al., 2021).

segments and an angle of 135° between segments were used. A suitable set of constitutive equations was applied, as discussed in Section 4.4.3.

4.4.3 CONSTITUTIVE EQUATIONS FOR THE DRYING MODELING CLOSURE

For the drying modeling closure, it is necessary to experimentally determine the drying rate (R_w) and the equilibrium solid moisture content (X_e), which are related through Equation (4.65):

$$R_w = k\left[X_s(t) - X_e\right] \tag{4.65}$$

where k is the drying constant (min⁻¹), which is commonly described by the Arrhenius expression as a function of the drying air temperature, as follows:

$$k = k_0 \exp\left(-\frac{E_a}{RT_f}\right), \tag{4.66}$$

where k_0 and E_a (activation energy) are the model parameters.

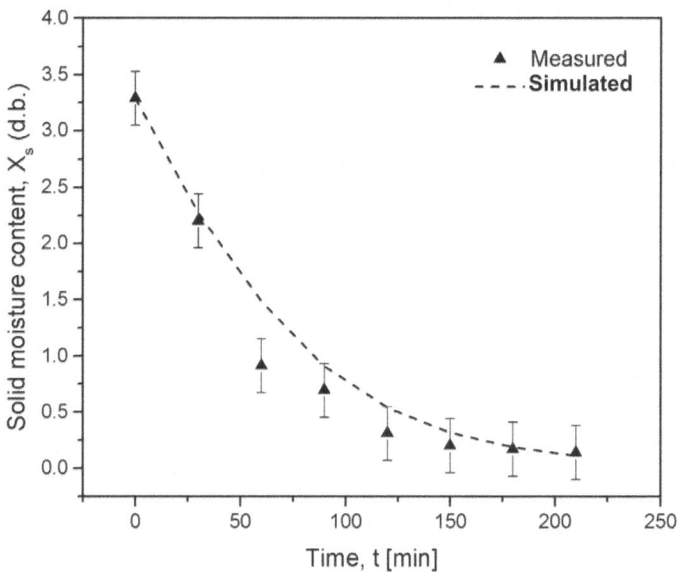

FIGURE 4.25 Experimental and simulated results of solid moisture content over time for a rotary dryer operating in the following conditions: mass flow rate of solids of 0.4 kg min^{-1}, inlet air velocity of 2.0 m s^{-1}, inlet air temperature of 155°C, initial solid moisture content of 3.2868 [kg H$_2$O kg^{-1} dry material], and inlet gas absolute humidity of 0.0692 [kg H$_2$O kg^{-1} dry air] (Perazzini et al., 2021).

TABLE 4.2
Equilibrium Solid Moisture Content Models

Model	Equation	
Henderson correlation	$$X_e = \left[\frac{\ln(1-RH)}{-aT_f} \right]^{1/b}$$	(4.67)
Henderson–Thompson correlation	$$X_e = \left[\frac{\ln(1-RH)}{-a(T_f+c)} \right]^{1/b}$$	(4.68)
Chung–Pfost correlation	$$X_e = -\frac{1}{b}\ln\left[\frac{(T_f+c)\ln(RH)}{-a} \right]$$	(4.69)
Chen–Clayton correlation	$$X_e = -\frac{1}{cT_f^d}\ln\left[\frac{\ln(RH)}{-aT_f^b} \right]$$	(4.70)
Modified Halsey correlation	$$X_e = \left[\frac{-\exp(aT_f+c)}{\ln(RH)} \right]^{1/b}$$	(4.71)

TABLE 4.3
Drying Kinetics Models

Model	Equation	
Lewis	$\dfrac{X_s - X_e}{X_0 - X_e} = \exp(-kt)$	(4.72)
Brooker	$\dfrac{X_s - X_e}{X_0 - X_e} = a \times \exp(-kt)$	(4.73)
Page	$\dfrac{X_s - X_e}{X_0 - X_e} = \exp(-kt^n)$	(4.74)
Overhults	$\dfrac{X_s - X_e}{X_0 - X_e} = \exp\left[-(kt)^n\right]$	(4.75)

Many drying kinetics and equilibrium moisture content models have been correlated with experimental data to obtain the drying constant k and the equilibrium solid moisture content X_e, respectively, which depend mostly on the physical characteristics of the solid material, the drying air temperature range, the solid temperature range, and the drying air humidity range (Iguaz et al., 2003; Arruda et al., 2009b; Silva et al., 2012; Souza et al., 2021; Perazzini et al., 2021). Tables 4.2 and 4.3 list some models commonly used in the literature to represent k and X_e as functions of the drying operating conditions, where RH is the drying air relative humidity, X_0 is the initial solid moisture content ($t = 0$), and a, b, c, d, and n are the model parameters.

Finally, the volumetric heat transfer coefficient $(U_v a)$ and the heat transfer coefficient for heat loss from the control volume (U_p) can be determined by experimental correlations, such as those shown in Equations (4.76) and (4.77), respectively (Fernandes et al., 2009; Arruda et al., 2009b; Perazzini, 2011; Silva et al., 2012; Souza et al., 2021).

$$U_v a = A_1 G_f^{A_2} G_s^{A_3} \tag{4.76}$$

$$U_p = A_4 G_f^{A_5} \tag{4.77}$$

where A_1, A_2, A_3, A_4, and A_5 are the model parameters.

REFERENCES

Agarwal, P.K., "Transport phenomena in multi-particle systems – II. Particle-fluid heat and mass transfer" *Chemical Engineering Science* 43 (1988): 2501–2510. https://doi.org/10.1016/0009-2509(88)85184-4.

Alder, B.J.; Wainwright, T.E. "Phase transition for a hard sphere system" *The Journal of Chemical Physics* 27 (1957): 1208–1209. https://doi.org/10.1063/1.1743957.

Arruda, E.B.; Lobato, F.S.; Assis, A.J.; Barrozo, M.A.S. "Modeling of fertilizer drying in roto-aerated and conventional rotary dryers" *Drying Technology* 27 (2009): 1192–1198. https://doi.org/10.1080/07373930903263129.

ASTM G194-08 "Standard test method for measuring rolling friction characteristics of a spherical shape on a flat horizontal plane", West Conshohocken, PA: ASTM International (2018). doi: 10.1520/G0194-08R18.

Bagnold, R.A., "Experiments on a gravity-free dispersion of large solids spheres in a Newtonian fluid under shear" *Proc. Roy. Soc.*, A225 (1954): 49–63. https://doi.org/10.1098/rspa.1954.0186.

Batista, J.N.M.; Santos, D.A.; Béttega, R. "Determination of the physical and interaction properties of sorghum grains: Application to computational fluid dynamics, discrete element method simulations of the fluid dynamics of a conical spouted bed" *Particuology* 54 (2021): 91–101. https://doi.org/10.1016/j.partic.2020.04.005.

Benedito, W.M.; Duarte, C.R.; Barrozo, M.A.S.; Santos, D.A. "An investigation of CFD simulations capability in treating non-spherical particle dynamics in a rotary drum" *Powder Technology* 332 (2018): 171–177. https://doi.org/10.1016/j.powtec.2018.03.067.

Benedito, W.M.; Duarte, C.R.; Barrozo, M.A.S.; Santos, D.A. "Cataracting-centrifuging transition investigation using nonspherical and spherical particles in a rotary drum through CFD simulations" *Particuology* 60 (2022): 48–60. https://doi.org/10.1016/j.partic.2021.03.012.

Box, M.J.; Hunter, W.G.; Hunter, J.S. *Statistics for Experimenters. An Introduction to Design, Data Analysis, and Model Building*, John Wiley and Sons, New York, 1978.

Caserta, A.J.; Navarro, H.A.; Cabezas-Gómez, L. "Damping coefficient and contact duration relations for continuous nonlinear spring-dashpot contact model in DEM" *Powder Technology* 302 (2016): 462–479. https://doi.org/10.1016/j.powtec.2016.07.032.

Chen, Y.; Jiang, P.; Xiong, T.; Wei, W.; Fang, Z.; Wang, B. "Drag and heat transfer coefficients for axisymmetric nonspherical particles: A LBM study" *Chemical Engineering Journal* 424 (2021): 130391. https://doi.org/10.1016/j.cej.2021.130391.

Coetzee, C. "Calibration of the discrete element method: Strategies for spherical and non-spherical particles" *Powder Technology* 364 (2020): 851–878. https://doi.org/10.1016/j.powtec.2020.01.076.

Coetzee, C.J. "Review: Calibration of the discrete element method" *Powder Technology* 310 (2017): 104–142. https://doi.org/10.1016/j.powtec.2017.01.015.

Cokljat, D.; Ivanov, V.A.; Sarasola, F.J.; Vasquez, S.A. "Multiphase k-epsilon models for unstructured meshes", in: *ASME 2000 Fluids Engineering Division Summer Meeting*, Boston, MA, (2000): 749–754.

Cundall, P.A.; Strack, O.D.L. "A discrete numerical model for granular assemblies" *Géotechnique* 29 (1979): 47–65. https://doi.org/10.1680/geot.1979.29.1.47.

Danby, M.; Shrimpton, J.; Palmer, M. "On the optimal numerical time integration for DEM using Hertzian force models" *Computers & Chemical Engineering* 58 (2013): 211–222. https://doi.org/10.1016/j.compchemeng.2013.06.018.

Derringer, G.; Suich, R. "Simultaneous optimization of several response variables" *Journal of Quality Technology* 12 (1980): 214–219. https://doi.org/10.1080/00224065.1980.11980968.

Di Maio, F.P.; Di Renzo, A. "Modelling particle contacts in distinct element simulations: Linear and non-linear approach" *Chemical Engineering Research and Design* 83 (2005): 1287–1297. https://doi.org/10.1205/cherd.05089.

Douglas, P.L.; Kwade, A.; Lee, P.L.; Mallick, S.K. "Simulation of a rotary dryer for sugar crystalline" *Drying Technology* 11 (1993): 129–155. https://doi.org/10.1080/07373939308916806.

Elghobashi, S.E.; Abou-Arab, T.W. "A two-equation turbulence model for two-phase flows" *Physics of Fluids* 26 (1983): 931–938. https://doi.org/10.1063/1.864243.

Ergun, S. "Fluid flow through packed columns" *Chemical Engineering Progress* 48 (1952): 89–94.

Fernandes, N.J.; Ataíde, C.H.; Barrozo, M.A.S. "Modeling and experimental study of hydrodynamic and drying characteristics of an industrial rotary dryer" *Brazilian Journal of Chemical Engineering* 26 (2009): 331–341. https://doi.org/10.1590/S0104-66322009000200010.

Gibilaro, L.G.; Di Felice, R.; Waldram, S.P. "Generalized friction factor and drag coefficient correlations for fluid-particle interactions" *Chemical Engineering Science* 40 (1985): 1817–1823. https://doi.org/10.1016/0009-2509(85)80116-0.

Gidaspow, D. *Multiphase Flow and Fluidization. Continuum and Kinetic Theory Descriptions*, Elsevier Science, San Diego, CA, 1994.

Hertz, H. "Über die Berührung fester elastischer Körper" *The Journal für die reine und angewandte Mathematik* 92 (1882): 156–171.

Huang, Z.; Wang, L.; Li, Y.; Zhou, Q. "Direct numerical simulation of flow and heat transfer in bidisperse gas-solid systems" *Chemical Engineering Science* 239 (2021): 116645. https://doi.org/10.1016/j.ces.2021.116645.

Huilin, L.; Gidaspow, D.; Bouillard, J.; Wentie, L. "Hydrodynamic simulation of gas-solid flow in a riser using kinetic theory of granular flow" *Chemical Engineering Journal* 95 (2003): 1–13. https://doi.org/10.1016/S1385-8947(03)00062-7.

Iguaz, A.; Esnoz, A.; Martínez, G.; López, A.; Virseda, P. "Mathematical modelling and simulation for the drying process of vegetable wholesale by-products in a rotary dryer" *Journal of Food Engineering* 59 (2003): 151–160. https://doi.org/10.1016/S0260-8774(02)00451-X.

Jasak, H. "Dynamic mesh handling in OpenFOAM", in: *47th AIAA Aerospace Sciences Meeting Including the New Horizons Forum and Aerospace Exposition*, Orlando, FL, 2009.

Jiang, Z.; Du, J.; Rieck, C.; Bück, A.; Tsotsas, E. "PTV experiments and DEM simulations of the coefficient of restitution for irregular particles impacting on horizontal substrates" *Powder Technology* 360 (2020): 352–365. https://doi.org/10.1016/j.powtec.2019.10.072.

Johnson, P.C.; Jackson, R. "Frictional-collisional constitutive relations for granular materials with application to plane shearing" *Journal of Fluid Mechanics* 176 (1987): 67–93. https://doi.org/10.1017/S0022112087000570.

Langston, P.A.; Tüzün, U.; Heyes, D.M. "Continuous potential discrete particle simulations of stress and velocity fields in hoppers: Transition from fluid to granular flow" *Chemical Engineering Science* 49 (1994): 1259–1275. https://doi.org/10.1016/0009-2509(94)85095-X.

Larachi, F.; Alix, C.; Grandjean, B.P.A.; Bernis, A. "Nu/Sh correlation for particle-liquid heat and mass transfer coefficients in trickle beds based on Péclet similarity" *Trans IChemE* 81 (2003): 689–694. https://doi.org/10.1205/026387603322150534.

Launder, B.E.; Spalding, D.B. "The numerical computation of turbulent flows" *Computer Methods in Applied Mechanics and Engineering* 3 (1974): 269–289. https://doi.org/10.1016/0045-7825(74)90029-2.

Lima, R.M.; Brandao, R.J.; Santos, R.L.; Duarte, C.R.; Barrozo, M.A.S. "Analysis of methodologies for determination of DEM input parameters" *Brazilian Journal of Chemical Engineering* 38 (2021): 287–296. https://doi.org/10.1007/s43153-021-00107-4.

Lu, G.; Third, J.R.; Müller, C.R. "Discrete element models for non-spherical particle systems: From theoretical developments to applications" *Chemical Engineering Science* 127 (2015): 425–465. https://doi.org/10.1016/j.ces.2014.11.050.

Luding, S. "Introduction to discrete element methods" *European Journal of Environmental and Civil Engineering* 12 (2008): 785–826. https://doi.org/10.1080/19648189.2008.9693050.

Lun, C.K.K.; Savage, S.B.; Jeffrey, D.J.; Chepurniy, N. "Kinetic theories for granular flow: Inelastic particles in coquette flow and singly inelastic particles in a general flow field" *Journal of Fluid Mechanics* 140 (1984): 223–256. https://doi.org/10.1017/S0022112084000586.

Machado, M.V.C.; Nascimento, S.M.; Duarte, C.R.; Barrozo, M.A.S. "Boundary conditions effects on the particle dynamic flow in a rotary drum with a single flight" *Powder Technology* 311 (2017): 341–349. https://doi.org/10.1016/j.powtec.2017.01.076.

Menter, F.R. "Two-equation eddy-viscosity turbulence models for engineering applications" *AIAA Journal* 32 (1994): 1598–1605. https://doi.org/10.2514/3.12149.

Mindlin, R.D.; Deresiewicz, H. "Elastic spheres in contact under varying oblique forces" *Journal of Applied Mechanics* 20 (1953): 327–344. https://doi.org/10.1115/1.4010702.

Nascimento, S.M.; Duarte, C.R.; Barrozo, M.A.S. "Analysis of the design loading in a flighted rotating drum using high rotational speeds" *Drying Technology* 36 (2018): 1200–1208. https://doi.org/10.1080/07373937.2017.1392972.

Nascimento, S.M.; Lima, R.M.; Brandão, R.J.; Santos, D.A.; Duarte, C.R.; Barrozo, M.A.S. "Comparison between the Eulerian (CFD) and the Lagrangian (DEM) approaches in the simulation of a flighted rotary drum" *Computational Particle Mechanics* (2021): https://doi.org/10.1007/s40571-021-00407-z.

Otsubo, M.; O'Sullivan, C.; Shire, T. "Empirical assessment of the critical time increment in explicit particulate discrete element method simulations" *Computers and Geotechnics* 86 (2017): 67–79. https://doi.org/10.1016/j.compgeo.2016.12.022.

Paulick, M.; Morgeneyer, M.; Kwade, A. "Review on the influence of elastic particle properties on DEM simulation results" *Powder Technology* 283 (2015): 66–76. https://doi.org/10.1016/j.powtec.2015.03.040.

Perazzini, H. Drying of citrus solid wastes in a rotary dryer, Master Dissertation, Federal University of São Carlos (2011), São Carlos, São Paulo, Brazil (in Portuguese).

Perazzini, H.; Perazzini, M.T.B.; Freire, F.B.; Freire, F.B.; Freire, J.T. "Modeling and cost analysis of drying of citrus residues as biomass in rotary dryer for bioenergy" *Renewable Energy* 175 (2021): 167–178. https://doi.org/10.1016/j.renene.2021.04.144.

Podlozhnyuk, A.; Pirker, S.; Kloss, C. "Efficient implementation of superquadric particles in Discrete Element Method within an open-source framework" *Computational Particle Mechanics* 4 (2017): 101–118. https://doi.org/10.1007/s40571-016-0131-6.

Qi, Z.; Yu, A.B. "A new correlation for heat transfer in particle-fluid beds" *International Journal of Heat and Mass Transfer* 181 (2021): 121844. https://doi.org/10.1016/j.ijheatmasstransfer.2021.121844.

Ranz, W.E.; Marshall, W.R. "Evaporation from drops" *Chemical Engineering Progress* 48 (1952): 141–146.

Santos, D.A.; Barrozo, M.A.S.; Duarte, C.R.; Weigler, F.; Mellmann, J. "Investigation of particle dynamics in a rotary drum by means of experiments and numerical simulations using DEM" *Advanced Powder Technology* 27 (2016): 692–703. https://doi.org/10.1016/j.apt.2016.02.027.

Schaeffer, G. "Instability in the evolution equations describing incompressible granular flow" *The Journal of Differential Equations* 66 (1987): 19–50. https://doi.org/10.1016/0022-0396(87)90038-6.

Schiller, L.; Naumann, Z. "A drag coefficient correlation" *Zeitschrift des Vereins Deutscher Ingenieure* 77 (1935): 318.

Silva, M.G.; Lira, T.S.; Arruda, E.B.; Murata, V.V.; Barrozo, M.A.S. "Modelling of fertilizer drying in a rotary dryer: Parametric sensitivity analysis" *Brazilian Journal of Chemical Engineering* 29 (2012): 359–369. https://doi.org/10.1590/S0104-66322012000200016.

Silveira, J.C.; Brandao, R.J.; Lima, R.M.; Machado, M.V.C.; Barrozo, M.A.S.; Duarte, C.R. "A fluid dynamic study of the active phase behavior in a rotary drum with flights of two and three segments" *Powder Technology* 368 (2020): 297–307. https://doi.org/10.1016/j.powtec.2020.04.051.

Silveira, J.C.; Lima, R.M.; Brandao, R.J.; Duarte, C.R.; Barrozo, M.A.S. "A study of the design and arrangement of flights in a rotary drum" *Powder Technology* 395 (2022): 195–206. https://doi.org/10.1016/j.powtec.2021.09.043.

Souza, G.F.M.V.; Avendano, P.S.; Francisquetti, M.C.C.; Ferreira, F.R.C.; Duarte, C.R.; Barrozo, M.A.S. "Modeling of heat and mass transfer in a non-conventional rotary dryer" *Applied Thermal Engineering* 182 (2021): 116118. https://doi.org/10.1016/j.applthermaleng.2020.116118.

Syamlal, M.; O'Brien, T.J. "Simulation of granular layer inversion in liquid fluidized beds" *International Journal of Multiphase Flow* 14 (1988): 473–481. https://doi.org/10.1016/0301-9322(88)90023-7.

Syamlal, M.; Rogers, W.; O'Brien, T.J. *MFIX Documentation: Volume 1, Theory Guide*, National Technical Information Service, Springfield, 1993.

Ullah, A.; Hong, K.; Gao, Y.; Gungor, A.; Zaman, M. "An overview of Eulerian CFD modeling and simulation of non-spherical biomass particles" *Renewable Energy* 141 (2019): 1054–1066. https://doi.org/10.1016/j.renene.2019.04.074.

Versteeg, H.K.; Malalasekera, W. *An Introduction to Computational Fluid Dynamics: The Finite Volume Method*, 2nd ed., Pearson Education, England, 2007.

Walton, O.R.; Braun, R.L. "Viscosity, granular-temperature, and stress calculations for shearing assemblies of inelastic, frictional disks" *Journal of Rheology* 30 (1986): 949–980. https://doi.org/10.1122/1.549893.

Wang, S.; Zhang, Q.; Ji, S. "GPU-based parallel algorithm for super-quadric discrete element method and its applications for non-spherical granular flows" *Advances in Engineering Software* 151 (2021b): 102931. https://doi.org/10.1016/j.advengsoft.2020.102931.

Wang, S.; Zhou, Z.; Ji, S. "Radial segregation of a Gaussian-dispersed mixture of superquadric particles in a horizontal rotating drum" *Powder Technology* 394 (2021a): 813–824. https://doi.org/10.1016/j.powtec.2021.09.012.

Wen, C.Y.; Yu, Y.H. "Mechanics of fluidization" *Chemical Engineering Progress Symposium Series* 62 (1966): 100–111.

Zhang, L.; Jiang, Z.; Mellmann, J.; Weigler, F.; Herz, F.; Bück, A.; Tsotsas, E. "Influence of the number of flights on the dilute phase ratio in flighted rotating drums by PTV measurements and DEM simulations" *Particuology* 56 (2021): 171–182. https://doi.org/10.1016/j.partic.2020.09.010.

Zhang, L.; Jiang, Z.; Weigler, F.; Herz, F.; Mellmann, J.; Tsotsas, E. "PTV measurement and DEM simulation of the particle motion in a flighted rotating drum" *Powder Technology* 363 (2020): 23–37. https://doi.org/10.1016/j.powtec.2019.12.035.

Zhong, H.; Lan, X.; Gao, J.; Zheng, Y.; Zhang, Z. "The difference between specularity coefficient of 1 and no-slip solid phase wall boundary conditions in CFD simulation of gas–solid fluidized beds" *Powder Technology* 286 (2015): 740–743. https://doi.org/10.1016/j.powtec.2015.08.055.

Index